FROM THE GROUND UP

The **Institute of Southeast Asian Studies (ISEAS)** was established as an autonomous organization in 1968. It is a regional centre dedicated to the study of socio-political, security and economic trends and developments in Southeast Asia and its wider geostrategic and economic environment. The Institute's research programmes are the Regional Economic Studies (RES, including ASEAN and APEC), Regional Strategic and Political Studies (RSPS), and Regional Social and Cultural Studies (RSCS).

ISEAS Publishing, an established academic press, has issued more than 2,000 books and journals. It is the largest scholarly publisher of research about Southeast Asia from within the region. ISEAS Publishing works with many other academic and trade publishers and distributors to disseminate important research and analyses from and about Southeast Asia to the rest of the world.

FROM THE GROUND UP

Perspectives on Post-Tsunami and Post-Conflict Aceh

Edited by
Patrick Daly • R. Michael Feener • Anthony Reid

ISEAS

INSTITUTE OF SOUTHEAST ASIAN STUDIES
SINGAPORE

First published in Singapore in 2012 by
ISEAS Publishing
Institute of Southeast Asian Studies
30 Heng Mui Keng Terrace
Pasir Panjang
Singapore 119614

E-mail: publish@iseas.edu.sg
Website: <http://bookshop.iseas.edu.sg>

The responsibility for facts and opinions in this publication rests exclusively with the authors and their interpretations do not necessarily reflect the views or the policy of the publisher or its supporters.

ISEAS Library Cataloguing-in-Publication Data

From the ground up : perspectives on post-tsunami and post-conflict Aceh / edited by Patrick Daly, R. Michael Feener, Anthony Reid.
1. Disaster relief—Indonesia—Aceh.
2. Tsunamis—Indonesia—Aceh.
3. Peace-building—Indonesia—Aceh.
4. Aceh (Indonesia)—Politics and government.
I. Daly, Patrick.
II. Feener, R. Michael.
III. Reid, Anthony, 1939–
HV555 I5F93   2012

ISBN 978-981-4345-19-4 (soft cover)
ISBN 978-981-4345-20-0 (e-book, PDF)

Cover photograph by Patrick Daly: Aceh, Indonesia, 2007. The spray painted words in the building reads: "It is better to stay here in our village".

Typeset by Superskill Graphics Pte Ltd
Printed in Singapore by

CONTENTS

PREFACE

The tsunami that struck a dozen countries around the Indian Ocean on 26 December 2004 evoked international sympathy on a scale beyond any previous natural disaster. The unprecedented media coverage and humanitarian response was prompted not only by dramatic images relayed from hand-held cameras and phones, but by the inclusion of "First World" victims in an essentially "Third World" catastrophe. Among the areas hit by the tsunami were popular beach resorts in southern Thailand and Sri Lanka; Europeans, Americans and Australians were among the Indonesians, Indians, Thais and Sri Lankans who perished in huge numbers. The international relief effort broke all records both in scale and diversity, with seven billion U.S. dollars donated from all over the world through public and private agencies for Sumatra alone.

The disbursement of those funds and the rebuilding of housing, infrastructure and economy posed major national and international challenges. Indonesian President Susilo Bambang Yudhoyono (SBY) welcomed an unprecedented international relief effort which brought thousands of government and private aid workers to Aceh, transforming it from isolated backwater to international hub. After some initial uncertainty, he sidestepped the Indonesian bureaucracy and took the unprecedented step of establishing the novel Agency for the Rehabilitation and Reconstruction of Aceh and Nias (known by its Indonesian initials, BRR). The head of BRR, Kuntoro Mangkusubroto, had complete autonomy to act, as a minister responsible directly to the president.

However, this was not simply a reconstruction effort. Aceh at that time was a war zone; Indonesia's military was engaged in a major operation to crush a separatist rebellion that had been simmering since 1976. Curiously, two other hotbeds of separatism and repression, southern Thailand and Sri Lanka, were also severely affected by the 2004 tsunami, but without any peace dividend. In Aceh, however, the scale of the disaster, in conjunction with some other factors detailed in this book, became part of the remarkable peace of 2005. Even though the funds had been donated for tsunami relief, any real reconstruction of Aceh had to consider the impact of the conflict on the well-being of the population, as well as on governance and administrative capacities. Regardless of the exact nature of the relationship between the conflict and the tsunami, timing dictated that processes of reconstruction, reintegration and development had to address both sets of dynamics.

During the exceptional period from 2005–09, both the reconstruction and peace processes in Aceh were highly international, in a province where conflict had virtually excluded foreigners for decades. The Helsinki peace agreement mandated that the ceasefire and disarmament would be monitored by the Aceh Monitoring Mission (AMM), headed by the European Community's Pieter Feith. The success of the peace process created the stable platform and security necessary for the longer-term reconstruction and development efforts. It was within this context that the BRR funded an initial International Conference of Aceh and Indian Ocean Studies (ICAIOS) in Banda Aceh on 24–27 February 2007. Its organization was entrusted to the Asia Research Institute (ARI) of the National University of Singapore, of which Anthony Reid was then Director, and to the other editors of this volume. The Indian Ocean context was intended to emphasize that Aceh's significance was not limited to Sumatra or Indonesia, but was enmeshed by geography, history and the tsunami in a much wider world.

Most of the chapters in this book derived from three of the six conference panels, devoted to seismology, geology and environmental issues; conflict resolution, peacemaking and democratization; and disaster relief and reconstruction. The other three panels dealt with history; Islamic law and society; and language, culture and society in Aceh. The bilingual discussions in Aceh during the conference generated great local interest; group discussion sessions ensured that Acehnese academics and intellectuals could debate with colleagues from around the world to evaluate the state of knowledge and the way forward towards a more open future. The relationships begun there have deepened and improved the chapters. Papers dealing with Acehnese history are assembled in a companion volume by the same editors.[1] These volumes serve the purpose not only of discussing some of the lessons of the Aceh reconstruction and peace processes, but also of maintaining critical links between Aceh and the international community after the initial tranches of aid expire. Both of these books were prepared in the hope that Aceh will continue to be engaged in wider national and international contexts, and that it will move forward from its isolated and traumatic recent past.

The editors would like to thank those who made the 2007 conference possible, notably Kuntoro Mangkusubroto and Heru Prasetjo of BRR, and the admirable Alyson Rozells of the Asia Research Institute. Saharah Abubakar and Joyce Zaide, also of ARI, made valuable contributions to the preparation of the manuscript.

Note

1. *Mapping the Acehnese Past*, ed. Michael Feener, Patrick Daly and Anthony Reid (Leiden: KITLV Press, 2011).

LIST OF FIGURES AND TABLES

Tables

THE CONTRIBUTORS

Leena Avonius received her Ph.D. in anthropology at Leiden University in the Netherlands in 2004. In 2005–06 she worked as EU observer and Reintegration Coordinator for the Aceh Monitoring Mission. She worked as a postdoctoral researcher at the University of Helsinki in Finland. She is currently the director of the International Center for Aceh and Indian Ocean Studies, Banda Aceh.

Stéphane Bernard is assistant professor at the School of International Development and Global Studies, University of Ottawa. His work deals mostly with agricultural and environmental issues in Southeast Asia, with a current focus on Borneo.

Ian Christoplos is a researcher at the Swedish University of Agricultural Sciences and an independent consultant. His research engagements focus on risk, recovery, rural development and agricultural services. He is the author of the Tsunami Evaluation Coalition's study *Links between Relief, Rehabilitation and Development in the Tsunami Response.*

Patrick Daly is a research fellow at the Asia Research Institute at the National University of Singapore. His research focuses on community-level responses to conflicts and natural disasters, most recently issues of reconstruction and recovery in Aceh, Indonesia, following the 2004 Asian Tsunami. He has conducted research as well as worked with local NGOs in Palestine, Cambodia, Philippines and Indonesia on issues related to reconstruction and reintegration post disaster/conflict.

Rodolphe De Koninck is professor of geography and Canada Chair of Asian Research at the University of Montreal. His work concerns primarily Southeast Asian agrarian and environmental issues. He has authored or co-authored 20 books, and published nearly 200 refereed articles, including *Agricultural Modernization, Poverty and Inequality,* co-authored with D.S. Gibbons and I. Hasan (1980), *Malay Peasants Coping with the World* (1992), *Deforestation in Viet Nam* (1999), *Singapour: la cité-État ambitieuse* (2006), *Malaysia: la dualité territoriale* (2007) and *Singapore: An Atlas of Perpetual Territorial Transformation,* co-authored with Julie Drolet and Marc Girard (2008).

R. Michael Feener is associate professor of history at the National University of Singapore and senior research fellow at the Asia Research Institute. Born in Salem, Massachusetts, he was trained in Islamic studies and foreign languages at Boston University, Cornell and the University of Chicago, as well as in Indonesia, Egypt and Yemen. His recent books include *Muslim Legal Thought in Modern Indonesia* (2007), *Islamic Law in Contemporary Indonesia: Ideas and Institutions* (2007) and *Islamic Connections: Muslim Societies in South and Southeast Asia* (2009).

Pieter Feith was appointed on 9 September 2005 by the Council of the European Union as head of mission for the EU-led Aceh Monitoring Mission (AMM), to help facilitate the implementation of the Memorandum of Understanding between the Government of Indonesia and the Free Aceh Movement (GAM), signed in Helsinki on 15 August 2005. He has spent his life in diplomatic service, with the Netherlands diplomatic service, NATO, UN and the European Union.

Wolfgang Fengler was the head economist of the World Bank in Jakarta, where he led the World Bank's Decentralization and Public Finance Team. He also led the World Bank's analytical work supporting Aceh. He was team leader for flagship reports such as "Aceh and Nias One Year after the Tsunami".

Daniel Fitzpatrick has written widely on land law and policy in the Third World, with a particular focus on recovery from disaster or conflict. He was the UN's land rights adviser in post-conflict East Timor (2000) and post-tsunami Aceh (2005–06). He is the author of the UN's *Guidelines on Land Programming after Natural Disasters* (in press). He has undertaken professional consultancies on law and development with AusAID, the Asian Development Bank, Oxfam International, the OECD, UNDP and UN-Habitat. His work with AusAID includes co-authoring the 2008 *Making Land Work* report for its Pacific Land Programme.

Marc Girard is a cartographer and GIS specialist in the Department of Geography, University of Montreal.

Ahya Ihsan is a research analyst in the Poverty Reduction and Economic Management Unit at World Bank Office Jakarta. Prior to joining the World Bank, he was a part-time staff member at the Provincial Planning Board (Bappeda) of Aceh-Indonesia. His research interests are public finance, decentralization and intergovernmental transfers, and reconstruction finance. He is currently a core team member in the Macro and Fiscal Policy team, analysing fiscal policy development and government expenditure in Indonesia. He has led several studies, including Aceh Public Expenditure Analysis (2006, 2008), designing the intergovernmental transfer formula of the special autonomy fund in Aceh, and reconstruction finance analysis after the 2004 Indian Ocean Tsunami.

Kai Kaiser is senior economist at the Public Sector Group, PREM. He is co-leader of the World Bank's Decentralization and Sub-National Regional Economics Thematic Group. Previously he was based in Jakarta, Indonesia, where he worked on post-tsunami reconstruction. He has extensive experience across all regions in intergovernmental fiscal reform issues

Saiful Mahdi is a Ph.D. candidate in Regional Sciences, Department of City and Regional Planning, Cornell University, Ithaca, New York. He is on study leave from his lecture position at Syiah Kuala University, Banda Aceh. He is a co-founder of The Aceh Institute, an independent research institute in Banda Aceh of which he is executive director.

Michael Morfit is senior vice-president at Partners for Democratic Change, a nonprofit organization that specializes in conflict resolution and dispute mediation in the developing world. He is adjunct professor at Georgetown University and the American University in Washington D.C., where he teaches courses on democracy, governance and development.

Yenny Rahmayati is writing her Ph.D. in Architecture at School of Design and Environment, National University of Singapore. Her research focuses on the socio-cultural impacts of disaster and reconstruction. She is the Director of the Aceh Heritage Community Foundation, which works on architectural and cultural heritage issues in Aceh. She was active in during the reconstruction process, working for several international aid organizations. She is a leading member of the Lestari Heritage Network, which promotes heritage issues and activities in Asia and the West Pacific.

Anthony Reid is a Southeast Asian historian at the Australian National University, after serving as founding director of the Center for Southeast Asian Studies at UCLA (1999–2002) and of the Asia Research Institute at the National University of Singapore (2002–07). His relevant and recent books include *The Contest for North Sumatra: Atjeh, the Netherlands and Britain, 1858–1898* (1969); *The Blood of the People: Revolution and the End of Traditional Rule in Northern Sumatra* (1979); *Southeast Asia in the Age of Commerce, 1450–1680*, 2 vols. (1988–93); *An Indonesian Frontier: Acehnese and Other Histories of Sumatra* (2004); *Imperial Alchemy: Nationalism and Political Identity in Southeast Asia* (2010), and *Verandah of Violence: The Historical Background of the Aceh Problem* (as editor; 2006).

Kerry Sieh is professor of geology and director of the Earth Observatory of Singapore at Nanyang Technological University. A former professor at the California Institute of Technology, his principal research interest is earthquake geology, which uses geological layers and landforms to understand the geometries of active faults, the earthquakes they generate, and the crustal structure their movements produce. More recently, he has investigated the multitude of active faults in Taiwan and figured out how their earthquakes are creating that mountainous island. He is currently exploring the earthquake geology of Myanmar. His principal research interest is the subduction megathrust that produced the devastating giant Sumatran earthquakes and Indian Ocean tsunamis of 2004 and 2005.

Rizal Sukma is currently deputy executive director at the Centre for Strategic and International Studies (CSIS), Jakarta. He received his Ph.D. in international relations from the London School of Economics and Political Science (LSE) in 1997. He has worked extensively on Southeast Asia's security issues, ASEAN, Indonesia's defence and foreign policy, military reform, Islam and politics, and domestic political changes in Indonesia.

John Telford is an Irish consultant specializing in international aid. In the last twenty-eight years he has worked as a practitioner, lecturer, analyst, evaluator, teacher and trainer worldwide, especially in Latin America, covering both natural disaster and conflict-related aid. Previously, he was a senior UN official, teacher, journalist, and national director of the Irish section of Amnesty International. He is one of the authors of the Tsunami Evaluation Coalition's *Joint Evaluation of the International Response to the Indian Ocean Tsunami.*

Treena Wu is a Ph.D. research fellow funded by the European Commission Marie Curie Fellowship at the Maastricht Graduate School of Governance, Maastricht University. Her research focuses on the design of social protection policy in Southeast Asia.

GLOSSARY AND ABBREVIATIONS

adat	customary law
AFEP	Aceh Forest and Environment Project
AMM	Aceh Monitoring Mission
aneuk yatim	orphan
ASEAN	Association of Southeast Asian Nations
BAKORNAS	Disaster Management Coordination Board
bantuan langsung tunai	direct cash aid
BAPPENAS	National Development Planning Board
BPN	*Badan Pertanahan Nasional*, Indonesian National Land Authority
BRA	*Badan Reintegrasi Aceh*, Aceh Reintegration Body
BRR	*Badan Rehabilitasi dan Rekonstruksi Aceh dan Nias*, Rehabilitation and Reconstruction Agency for Aceh and Nias
camat	head of a *kecamatan*, sub-district
CDA	Community Driven Adjudication
CFAN	Coordination Forum for Aceh and Nias
CMI	Crisis Management Initiative
CoHA	Cessation of Hostilities Agreement, signed by GAM and Indonesian government, 9 December 2003
COSA	Commission on Security Arrangements
DAD	Development Assistance Database
desa	Indonesian administrative unit at village level

DFID	Department for International Development
diyat	compensation given by the provincial authorities to the family of persons killed in the conflict
DPR	Indonesian National Parliament
DDR	disarmament, demobilization and reintegration
DRR	Disaster Risk Reduction
EC	European Commission
ERRA	Pakistan's Earthquake Reconstruction and Rehabilitation Agency
ESDP	European Security and Defense Policy
EU	European Union
fakir miskin	poor person
FOREC	*Fondo para la Reconstrucción y el Desarrollo Social del Eje Cafetero*, Agency established to coordination post-earthquake reconstruction in Colombia, 1999
Forum Bersama Damaia	Joint Peace Forum
GAM	*Gerakan Aceh Merdeka*, Free Aceh Movement
gampong	Acehnese village
gedung sosial	community hall, social building
GHD	Good Humanitarian Donorship
GLNP	Gunung Leuser National Park
GoI	Government of Indonesia
gotong royong	Indonesian concept of self-help and community involvement
GPS	Global Positioning System
gunung	mountain
harga mati	fixed price, red line, bottom line
HDC	Henry Dunant Centre
HRC	human rights court
hukum	law
ibu	mother
IDLO	International Development Law Organisation
IDP	Internally displaced person
IFES	International Foundation for Electoral Systems
IFI	International Financial Institutions
ilmu	knowledge
imam meunasah	*gampong*-level Islamic religious leader
IMP	Initial Monitoring Presence

INGO	International Non-governmental Organization
IOM	International Organization of Migration
ISEAS	Institute of Southeast Asian Studies
kabupaten	Indonesian government administrative unit at the regency level
kawom	kinship
KDP	*Kecamatan* Development Programme
keadilan	justice
kecamatan	subdistrict, Indonesian government administrative unit
kelurahan	Indonesian administrative unit at village level
kepala desa	village head
keucik	head of an Acehnese *gampong*
KPK	*Koalisi Pengungkapan Kebenaran*, Coalition for Truth Recovery
LRRD	Linking Relief, Rehabilitation and Development
LIF	Leuser International Foundation
LoGA	Law on Governing Aceh
LOGICA	Local Governance and Infrastructure for Communities in Aceh Project, Australia
lurah	head of a *kelurahan*
Mahkamah Syariah	Syariah Court
MDTF	Multi-Donor Trust Fund
MDG	Millennium Development Goals
merantau	to migrate in pursuit of economic fortune
meudagang	to trade
MoU	Helsinki Memorandum of Understanding
MPR	*Majalis Rakyat Papua*, Papuan People's Assembly
mupakat	agreement, consensus
musafir	traveller
NAD	*Nanggroe Aceh Darussalam*
NATO	North Atlantic Treaty Organization
NGO	Non-Governmental Organization
NKRI	*Negara Kesatuan Republik Indonesia*, principle of the territorial integrity of the Indonesian state
Operasi Terpadu	Military operation by the TNI to eradicate GAM — May 2003

OPM	*Organisasi Papua Merdeka* — Free Papua Organization
Patah	broken
PEFA	Public Expenditure & Financial Management Accountability
pendobrak	batterer, demolisher
pengungsi	refugee, IDP
perantauan	time spent on *merantau*
PFM	Public Financial Management
PFMA	Public Financial Management & Accountability
PIU	project implementation units
POLRI	Indonesian National Police
Posko	*Pos Komand*, centre for distribution of aid in Aceh
PRSP	Poverty Reduction Strategy Papers
RALAS	Reconstruction of Land Administration Systems in Aceh and Nias
rapat	meeting
reformasi	reform
saudara seperjuangan	Comrade in struggle
sayam	settle, pacify
SBY	Indonesian President Susilo Bambang Yudoyono
Sida	Swedish International Development Cooperation Agency
silaturrahmi	Muslim communal sentiment
suloh	Conflict resolution, peace settlement
SUSENAS	Indonesian National Socio-economic Survey
tawakkal	submission to and trust in God
TEC	Tsunami Evaluation Coalition
TKI	*Tenaga Kerja Indonesia*, Indonesian labour migrants
TKW	*Tenaga Kerja Wanita*, female Indonesia labour migrants
TLC	Temporary Living Centre
TNI	*Tentara Nasional Indonesia*, Indonesian Armed Forces
TRC	Truth and Reconciliation Commission
TRSH	Tropical Forest Heritage of Sumatra

tuha peut	village elders
uleebalang	Aceh territorial chief
UNCHS	United Nations Centre for Human Settlements
UNDP	United National Development Programme
UNHCR	United Nations High Commission for Refugees
UNIFEM	United Nations Development Fund for Women
UNIMS	United Nations Information Management System
UN-OCHA	United Nations Office for the Coordination of Humanitarian Affairs
UNORC	United Nations Recovery Coordinator for Aceh and Nias
UNTAET	United Nations Transitional Administration for East Timor
UUPA	*Undang Undang Pemerintahan Aceh*, 2006 Basic Law on the Governance of Aceh
waki	Representative, delegate

INTRODUCTION
Unpacking the Challenges of Post-2004 Aceh

Patrick Daly

How can we achieve post-disaster reconstruction and development that both rebuilds and protects people from potential loss in future catastrophes? How can we nurture a peace that assuages previous grievances and reduces the possibilities for renewed hostilities between parties with a long history of antagonism? These have been two of the main challenges facing Aceh following the 2004 Indian Ocean tsunami and the 2005 Helsinki Peace Accords that ended hostilities between the Gerakan Aceh Merdeka (GAM — Free Aceh Movement) and the Indonesian Government. Such questions are clearly important to the people of Aceh, who have experienced decades of conflict and isolation, the sudden devastation of the tsunami, and the painful and drawn-out process of mourning and rebuilding. The future of Aceh has been dramatically transformed by events since December 2004, and it will take years, if not decades, for things to stabilize.

The above questions are of immense importance to the wider international community. The experiences of Aceh will undoubtedly influence the texture and outcome of both future post-disaster responses and peace processes around the world. A large cadre of humanitarian aid workers, people involved in conflict resolution, and reconstruction and development advisers have already begun to bring their experiences from Aceh with them to their next posting or assignment. A whole generation of NGO staff, policy-makers and academics has been influenced by what happened around the Indian Ocean in the wake of the tsunami. The efforts in Aceh — both conflict and tsunami related — are extensively well documented and relatively transparent, opening possibilities for the kind of in-depth research and appraisal that are often not possible in the aftermath of large-scale trauma. The sheer amount of resources that was poured into the region to deal with the conflict and rebuild the shattered lives of victims of war and disaster warrants — indeed obligates — a comprehensive reflection in which previous standards are questioned and new knowledge generated.

This new knowledge should not be restricted to merely practical "lessons learned" or generic solutions that can be automatically applied during the next major crisis. Organizations, both governments and NGOs, have already put great efforts into compiling assessments of their actions in Aceh. These are widely available; several of the contributors to this volume have been deeply involved in the production of such analyses. While many of these are certainly of great value and necessary to the quality control and oversight processes for such organizations, they come with considerable limitations as well. There are pressures in the policy world that place serious time constraints on such assessments; many organizations typically do not have the mandate to engage in the long-term study necessary to more fully understand the implications and consequences of their work. Furthermore, it is hard to expect truly unbiased research and critiques on specific organizations and the wider humanitarian, reconstruction and development fields from people who are deeply implicated in, or funded, by such organizations.

Unfortunately, it is also typically the case that people with years of valuable on-the-ground experience working with relief, aid and development organizations are not encouraged to approach their work from a truly critical perspective, and are almost never given the chance to properly digest their experiences and distil what they have learned before moving on to the next "situation". Just as many of the early arrivals in Aceh were schooled in East Timor or the Balkans, many of those who have spent the past several years engaged around the Indian Ocean will inevitably move on to Afghanistan, Darfur, Haiti, Chile or wherever the next emergency arises that garners global attention. This rotation of talent leads to a fragmented view of the complex and long-term processes necessary for reconstituting shattered communities and societies. A brief conversation with staff at any large NGO will make it clear that many intelligent and experienced people have a wealth of insights that never formally see the light of day because of structural and institutional constraints.

It is also very difficult to get solid, meaningful output from academics that can be usefully factored into practical discussions of post-conflict and post-disaster situations. Academics are burdened with their own sets of institutional constraints. Unlike their counterparts in the policy and NGO worlds, very few academics from any discipline have the opportunity to spend enough time on the ground to develop deeper insights into the underlying dynamics of post-trauma situations. While distance and lack of affiliation lend a different perspective, arguably a useful and necessary one, they also reduce sustained, everyday engagement with the situations in question. Armchair reflections, drawing heavily upon NGO progress reports and supplemented by brief field visits, do not automatically give birth to profound new insights. Additionally, most researchers employed by academic institutions do not have to make the difficult decisions under less than optimal conditions that practitioners do.

The fragmented nature of academia leads to highly specialized foci on selected aspects of reconstruction and post-conflict situations. It is difficult to pull together and sustain the types of multidisciplinary partnerships that are necessary to foster more holistic research on what are incredibly complex problems. Furthermore, results from research often take years to pass through peer review and publication processes, and end up in journals or edited volumes that are rarely read by practitioners, or even other academics from different disciplines who work on the same broad problems. This can be easily verified by taking a quick look at the bookshelves of NGO workers on location which will often hold a handful of books deemed essential for a long stretch in the field. Rather than finding collections of articles or books by academics, one is much more likely to encounter a local language phrasebook, a Lonely Planet guidebook, maybe Naomi Klein's *The Shock Doctrine* and, in the case of Aceh, Anthony Reid's *Verandah of Violence*.

It is clear that our understanding of post-conflict and post-disaster situations faces serious institutional limits from all sides that need to be overcome. One way to do so is to foster more substantial interaction between practitioners and academics, involving international, national and "local" parties. This was part of the logic behind the First International Conference for Aceh and Indian Ocean Studies, held in Banda Aceh in February 2007, which served as the starting point for this volume. Additionally, it is important to find ways to bring together the valuable experiences and perspectives of members of both of these communities in a mutually constructive dialogue. Unfortunately, the standard formats within which each publishes does not readily support such efforts. Major organizations want pragmatic and focused assessments that can be easily translated into practice, while academic journals want contextualized and often more abstract discussions of the ideas. The end result is that the circles of academics and professionals working on the same issues are often largely distinct and disconnected.

Our aim with this volume is to provide a venue for the types of dialogue that are often lacking. The contributing authors have been asked to draw upon their varied experiences in Aceh to unpack some of the fundamental concepts underlying community recovery, reconciliation and governance, to name just a few. During the editorial process we respected the different styles and standards that the authors brought, and gave latitude for very different kinds of contributions. These range from academic research papers to broader policy assessments and personal and professional reflections. To begin to understand what has happened in Aceh since 2004, it is necessary to engage with all of these voices and perspectives.

CONTENTS OF THE VOLUME

The volume leads off with a chapter by Kerry Sieh, a leading professor of geology, who outlines the physical processes that resulted in the earthquake and tsunami. This gives the volume its temporal range, and sets the stage for the subsequent discussions of the post-tsunami reconstruction. Sieh goes to lengths to point out the continuing vulnerability in the region to future large-scale seismic events. His work in the region aims not just to produce scientific data about the nature of instability in the region, but also includes a significant outreach programme to inform people at the community level about the results of his team's research. His call for more integration of scientific and academic research with programmes of outreach is a necessary step to ensure that results and valuable information are made available to the people to whom it most matters, and to build up disaster mitigation efforts.

Following this initial chapter, the book is divided into two sections. The first section looks at the post-disaster situation. While it would be almost impossible to cover all of the relevant sectors, given the scale and complexity of operations in Aceh, the chapters were chosen to give a range of perspectives on some of the key facets of post-tsunami Aceh. Unfortunately, there are important issues not included in this book, such as reflections on the economic processes brought about by the relief and reconstruction efforts, and the outcome of the democratic political process. Given these recognized limitations, we hope that this book will make a solid contribution to what will become a vibrant and expansive body of literature.

Part I: Reconstruction Efforts

John Telford, an aid and development consultant and one of the authors of the Tsunami Evaluation Coalition (TEC) Report, provides an overview of some of the key findings and recommendations of the TEC, focusing on issues that are faced in many large-scale humanitarian aid missions. He discusses funding, illustrating how the unusual dynamics of funding in Aceh led to considerable overlap in projects, as well as to organizations far exceeding their sectoral or regional expertise. If the response to the tsunami is going to form the blueprint for how the international community engages with future major catastrophes, it is critical to understand core issues of feasibility and standards when large numbers of organizations get involved. In a theme which surfaces in a number of the contributions to this volume, Telford stresses the need for aid efforts to be more centred around and run by people in affected areas, which in spite of constant rhetoric is often far from the case in practice.

This next contribution draws heavily on the experiences of the TEC. Christoplos and Wu, both academics with extensive consulting and project evaluation experience, turn their attention on the Links between Relief, Rehabilitation and Development (LRRD), which has become a major element in post-conflict and post-disaster situations. As part of their chapter, they present a nuanced argument for the usefulness of having a more integrated reconstruction and development policy, and discuss some of the institutional shortcomings that disrupted the effectiveness of LRRD in Aceh. One of their main conclusions deals with the need to better ground LRRD efforts within local institutions to ensure suitability and sustainability. In what has become a standard part of post-disaster recovery vocabulary, LRRD emphasizes the need to focus not just on immediate solutions, but also on approaching underlying vulnerabilities to mitigate future disasters.

The next chapter, by Wolfgang Fengler, Ahya Ihsan and Kai Kaiser, looks at the mechanics of finance for post-tsunami operations from a more technical perspective. Given the scale of the budgets, and the serious reservations about accountability and transparency in Aceh, a number of institutions and oversight boards, such as the Multi-Donor Trust Fund, were put into place to monitor and attest to the distribution of donations and aid. This chapter brings a wealth of practical experience, obtained while Fengler was the head economist of the World Bank's Jakarta office, dealing with large-scale reconstruction finance, and it discusses some of the key dynamics in managing the diverse streams of funding that fuelled the recovery efforts around the Indian Ocean world. It also provides a useful starting point for better understanding how post-conflict and post-disaster reconstruction operations are funded and managed, while stressing the functional and budgetary differences between emergency humanitarian situations and longer-term development projects.

Daniel Fitzpatrick, drawing on his years of experience as an Indonesian land law expert and legal academic, contributes a chapter that looks at issues of land ownership and titling following the tsunami. Many basic aspects of reconstruction are predicated on having a workable plan for formalizing land ownership in cases where such issues are unclear. This complex process, led by the RALAS initiative of which Fitzpatrick was part, needed to contend with informal and customary notions of land ownership, Islamic legal guidelines, destruction and lack of records, and complex webs of inheritance. Additionally, it was necessary to reconcile such disparate notions of land ownership with the expectations and requirements of donor agencies, which included explicit consideration for issues of gender and inclusiveness. Interestingly, his research indicates that an emphasis upon formal inclusion of "gender-sensitive" programmes with regards to land titling did not always have the intended results. In fact, he suggests that in some cases such programmes excluded the very people they were designed to include.

Saiful Mahdi's paper, drawing on his experience as both an academic and prominent figure in Acehnese civil society, looks at how mobility has been used as a strategy by the Acehnese in response to both the conflict and the tsunami. During heightened times of conflict, and in the immediate post-tsunami period, internally displaced persons (IDPs) moved widely around Aceh, This chapter, grounded in a deep understanding of Acehnese cultural practices, emphasizes the cultural orientations of Acehnese towards movement, travel and hospitality, and the social capital implicit within village networks. He asks critical questions about the impact of aid efforts upon Acehnese responses, and suggests that many of the organizations involved in post-tsunami work neglected to adequately factor in the communal strengths and cultural attributes of the Acehnese.

Finally, the chapter by Rodolphe De Koninck, Stephane Bernard and Marc Girard gives us a deeper historical perspective on the nature of land change in Aceh. Their analysis of maps of forest cover and land use make it clear that there has been a definite increase in the rates of deforestation over the past several decades. Perhaps ironically, as they point out, the ending of the conflict opened up new possibilities for logging, as previously inaccessible lands became more open. Furthermore, efforts to connect disparate areas of Aceh also grant access to the deeply forested interior regions, facilitating logging operations. The forests and biodiversity of Aceh might be one of the most important and intact ecosystems in all of Southeast Asia; pressures of expanding agriculture and harvesting timber need to be balanced against the long-term social and economic importance of preserving Aceh's forests.

Part II: Conflict Resolution

The second section of the book looks at the peace process that ended the long-standing conflict between GAM and the Indonesian Government. Given the chronology of events in Aceh, it is only natural that the peace process and post-tsunami response have become conflated both in discussion and to some degree in the administrative mandates of government agencies and NGOs. While this connection does have some validity, and the two will remain linked in popular imagination, it is important to see them both as very different sets of processes, and to more fully understand the unique dynamics of each, as well as the areas of genuine overlap. All the contributing authors in this section have been deeply and, in some cases, personally involved in the conflict and peace process.

Michael Morfitt, in a very well-documented chapter, starts the section with a comprehensive discussion of how the peace process played out. Using his access to the key figures involved, he charts a narrative that began well before the tsunami, spanning several presidential administrations in post-New Order Indonesia. In addition to providing a very useful overview of the steps leading up to the signing of the Helsinki Accords in 2005, he clearly maps out the main actions and stances of all the pivotal players involved, ranging from the Indonesian President and Vice President and GAM leaders, to professional and "amateur" conflict resolution personnel. The story of peace in Aceh should not be subordinated to a secondary impact of the tsunami, but rather be seen for what it took: immense amounts of concerted efforts by various actors, bold decisions and stances by the political leadership on both sides of the conflict, and deep and engaged support from the international community. Finally, this chapter touches on the broader implications of the peace process for Indonesia, and for wider reforms within the country.

This chapter is nicely followed up by a more personal account by Pieter Feith, who oversaw the Aceh Monitoring Mission, charged with overseeing the implementation of the Helsinki Accords. Feith, an established figure in the conflict resolution field, emphasizes some of the key factors that so far have allowed the peace process to proceed with relative success. He provides a very different perspective on the situation and expands on the steps needed to ensure the successful maintenance of an enduring peace. While peace in

Aceh never could have occurred without the actions of the Acehnese and Indonesian leadership, it is equally unlikely that the implementation of the Helsinki Accords would have progressed smoothly in the absence of a firm and effective monitoring team. The presence of a third party that both sides could trust, and that could also bring the weight of EU financial, political and practical assistance to bear, has been a major part of the post-conflict story in Aceh.

Leena Avonius's chapter takes a more from-the-ground-up approach, looking at important issues of justice in post-conflict situations. It provides a theoretical overview of different forms of justice prevalent within the conversation of post-conflict situations, such as human rights courts and truth and reconciliation commissions. Additionally, drawing upon extensive personal experience in Aceh, Avonius makes a strong case for the need to better appreciate and work within Acehnese constructs and understandings of justice. As there are a number of different levels of authority within Aceh, including "official" government institutions, Islamic Sharia courts and village-level customary laws, it is essential to find a balance that is both broadly understood and applicable, while also sensitive to Acehnese customs, practices and institutional frameworks.

Finally, Rizal Sukma discusses the vast challenges of managing the peace in Aceh. As the international spotlight shifts from Aceh, the burden is increasing on local actors to assume full and unfettered responsibility for building a peaceful future for Aceh and a successful relationship between Aceh and Jakarta. This chapter presents a important Indonesian perspective on the events surrounding the peace process, as well as a focused perspective on the issues that still need to be resolved to ensure the area remains stable. Critical are continued improvements in the capacities for governance in Aceh, as well as a broader agenda of economic development throughout Aceh, in particular in areas that were not directly involved in the tsunami but nonetheless remain mired in poverty.

Overall, the two sections bring in a range of voices, all from people who have been deeply involved in various facets of the post-disaster and post-conflict processes in Aceh. These voices represent Indonesian, Acehnese and international parties, as well as academics, NGO workers and policy-makers. Some of the perspectives deal with macro-level institutional issues, while others focus more on smaller-scale issues. One of the threads that ties both together is the repeated emphasis within the chapters for post-disaster and post-conflict interventions to be firmly in line with "local" processes to maximize "local" ownership of such endeavours. This ensures a higher level of investment by people on the ground in Aceh to take responsibility for core decision-making and implementation, which leads to capacity creation and development. Furthermore, it is essential to long-term sustainability that all programmes fit within the blueprints that the Acehnese have for their lives post-2004. With the conflict over and communities moving forward from the trauma of the tsunami, Aceh has greater scope now to shape its future than at any other point in its recent history.

1

THE SUNDA MEGATHRUST
Past, Present and Future

Kerry Sieh

After lying dormant for about a thousand years, the sudden slippage of a 1,600-km long section of the Sunda megathrust fault caused uplift of the seafloor between the Indonesian island of Sumatra and Myanmar, resulting in a great earthquake and the horrific Indian Ocean tsunami of 2004. Three months later and just to the south, the slippage of a 350-km length of the same megathrust beneath Simeulue and Nias islands caused another destructive great earthquake and a lesser tsunami. While research indicates that it may be several hundred years before these areas experience another catastrophic earthquake, there are other highly populated areas along the Sunda fault that are vulnerable. It is imperative that we better understand the tectonic processes at play in the region, so that informed steps can be taken to promote disaster mitigation programmes, and avoid the scale of casualties caused by the 2004 earthquake and tsunami.

This chapter first provides a basic overview of what happened on 26 December 2004 from a geological perspective. This is in part to better contextualize the chapters in this volume that deal with the aftermath of the tsunami and also to situate my ongoing research in the region. I then discuss possibilities for future massive seismic events in the Indian Ocean region, drawing on several years of research. Finally, I make suggestions about how best to utilize this scientific knowledge to protect communities living in vulnerable coastal areas.

Similar future losses from earthquakes and tsunamis in South and Southeast Asia could, in theory, be substantially reduced. However, achieving this goal would require forging a strong chain that links knowledge of why, when and where these events will occur to people's everyday lives. The post-mortem of the 2004 disaster makes clear that the most important links in this chain are recognition and characterization of hazards through scientific research, then public education, emergency response preparedness, and improvement of infrastructural resilience.[1]

EARTHQUAKE AND TSUNAMI BASICS

Most great earthquakes occur at subduction zones — those zones of convergence between the Earth's tectonic plates where one is slowly sliding under the other. The contact surfaces between the two plates are known as megathrust faults. These resemble the thrust faults that are found, for example, under the cities of Los Angeles and Tehran, but are vastly larger. The Sunda megathrust runs south from Bangladesh, curving around the western and southern flanks of Sumatra, Java, Bali and eastern Indonesia to northwestern Australia — altogether a length of about 5,500 km. Other Asian megathrusts exist offshore from the Philippines, Taiwan, Japan and southeastern China. The biggest on-land megathrust runs from Pakistan through India and Nepal for a distance of 2,500 km along the southern side of the Himalayan mountain range.

Megathrusts commonly run from deep trenches on the ocean floor under the margins of continents. The fact that they lie underwater introduces a second hazard beyond the shaking caused by the earthquake itself: the rupture may suddenly displace a large volume of the overlying ocean, thereby triggering a tsunami. This is a wave that radiates out from the site of the strongest shaking, rapidly crosses the open ocean, and comes ashore tens to thousands of kilometres away as a series of waves and surges that can be metres or even tens of metres in height. It is ironic that at any one location, great subduction earthquakes and tsunamis occur at such long intervals that there is seldom any collective memory of previous events or alertness to future potential hazards.

Figure 1.1 shows the basic mechanisms of megathrust earthquakes and how they produce tsunamis. At the point of contact between the two plates, one plate (the Indian and Australian plates in this case) subducts beneath the other plate (the Sunda plate). The plane of contact between the plates is the megathrust, a gently sloping surface that descends from a deep ocean trench for several hundred kilometres into the Earth. Over the centuries between earthquakes, this megathrust remains locked, so the relative motion between

FIGURE 1.1

Idealized cross-section through the Sumatran plate boundary shows the accumulation and relief of strains associated with subduction. (a) Relationship of subducting plate (left) to overriding plate (right). The thick red line indicates the locked part of the megathrust between the two plates. (b) Since the megathrust is locked along this shallow portion, the over-riding block is squeezed and dragged downwards in the decades to centuries, leading up to a large earthquake. (c) Sudden relief of strains accumulated over centuries results in a large earthquake, uplift of the islands and a tsunami.

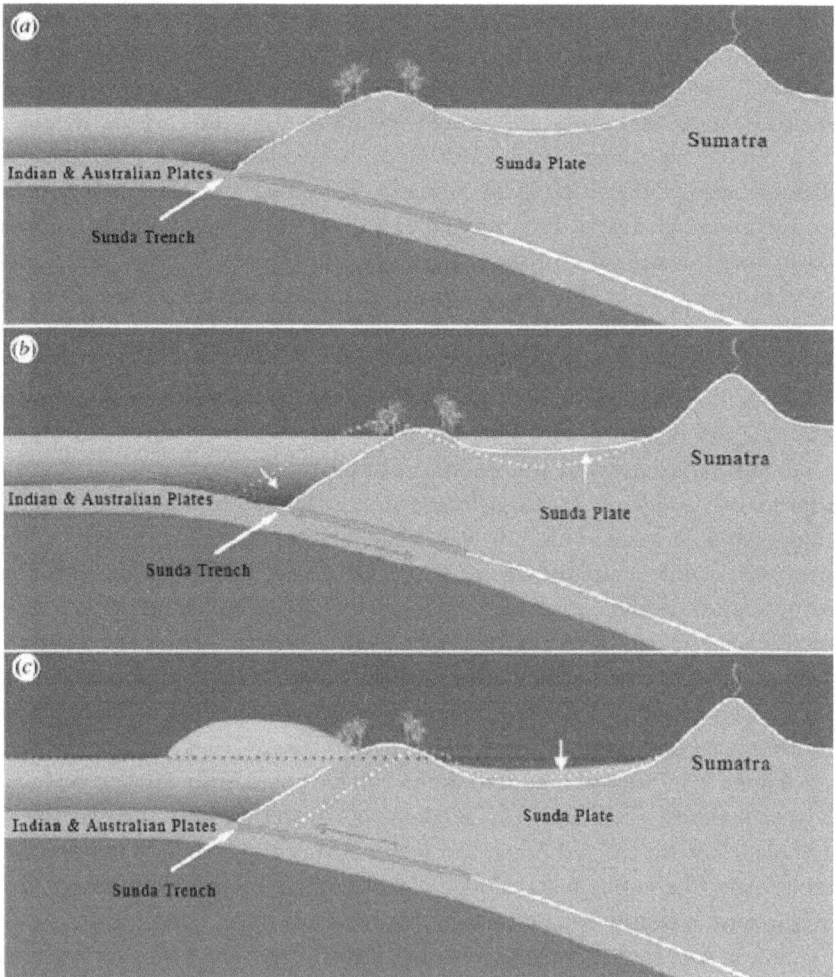

the two plates expresses itself not as a movement at the interface itself, but as a gradually increasing strain or deformation of the surrounding Earth's crust. Specifically, the advance of the subducting Indian and Australian plates causes the overlying Sunda plate to shorten and bow downwards in the region above the megathrust; it thus accumulates energy like a compressed spring or diving board (Figure 1.1b). When the accumulating stresses exceed the ability of the interface to withstand them, a rupture occurs; the Indian and Australian plates lurch forwards and downwards (by up to 10 metres in case of the 2005 earthquake), and the Sunda plate lurches back to its original, "relaxed" position (Figure 1.1c), dropping back to its original elevation. The lurching motion of the Sunda plate delivers a "kick" to the overlying ocean, thus triggering a tsunami.

PAST GREAT SUMATRAN EARTHQUAKES AND TSUNAMIS

2004 and 2005

The Sunda megathrust is the plane of contact formed as the Indian-Australian oceanic plates descend beneath the Sunda plate; the two descending plates are moving north–northeast with respect to the Sunda plate at a rate of about 50 mm per year (Figure 1.2). It was the rupture of a 1,600-km length of the megathrust that caused the magnitude (M) 9.2 earthquake of 26 December 2004. Tens of metres of sudden slip relieved centuries of slowly accumulating strain across the plate boundary. Movement of GPS monuments and the uplift and subsidence of corals show that the slip on the megathrust ranged to as high as about 20 m off the shores of Aceh and the Nicobar Islands.[2] The uplift of the seafloor caused by the slippages were as high as about 6 m. Measurement of uplifted coral reefs on Simeulue Island, above the southern end of the rupture (Figure 1.3) showed that the megathrust beneath that part of the island had slipped about 10 m. It was these uplifts that caused the great tsunami.

A second great earthquake occurred three months later, on 28 March 2005. In this M 8.7 event, the rupture of the subduction megathrust extended southwards an additional 350 km beyond the southern end of the 2004 rupture.[3] Once again, the pattern of movement of GPS monuments and corals allowed us to determine the length, depth and slippage on the megathrust. The combined length of the 2004 and 2005 ruptures is enormous — about 1,900 km. This is roughly the distance from Kuala Lumpur to Bali or from Singapore to Hanoi.

FIGURE 1.2

Setting and sources of the great 2004, 2005 and earlier earthquakes.
The pink patches overlie those sections of the megathrust that have failed
during large earthquakes.

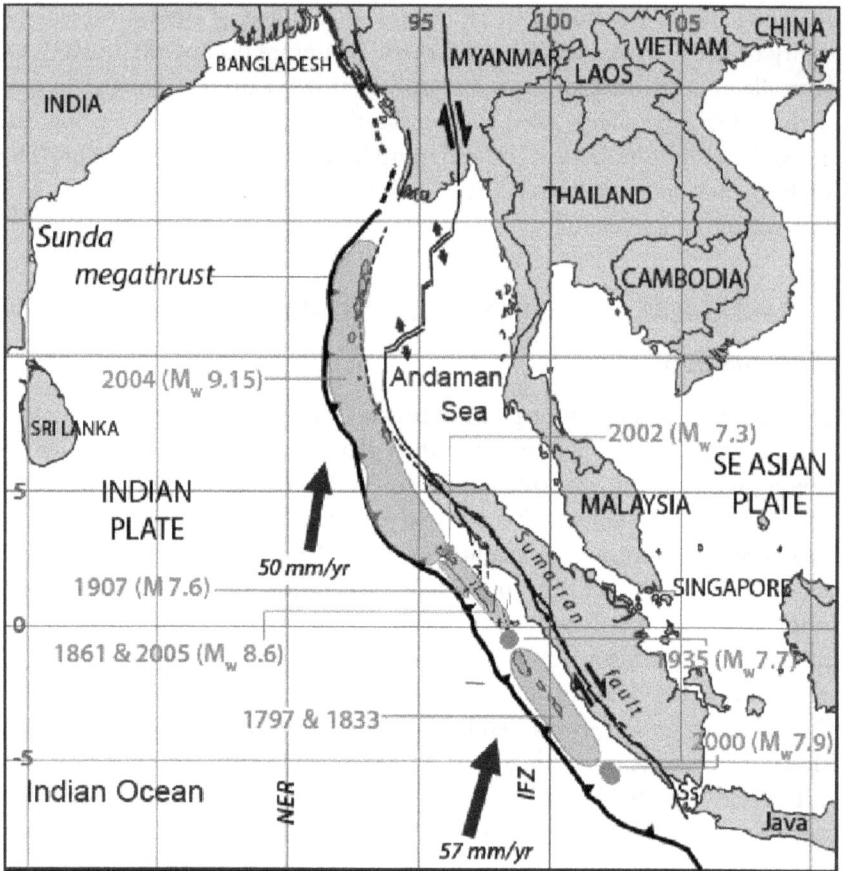

1797 and 1833

There are historical accounts of great earthquakes on sectors of the megathrust
to the south of the 2004 and 2005 events, but the accounts are too sparse to
tell us much about these large ancient earthquakes. Fortunately, however, we
have been able to use corals to characterize in detail these events of 1797 and

FIGURE 1.3

Aerial photograph of the western tip of Simeulue Island shows uplift of the fringing coral reef, evidence that the megathrust 25 km below the island slipped about 10 m during the 2004 earthquake.

1833.[4] Moreover, we have used modern GPS geodesy to measure the current accumulations of strain building towards the next big megathrust failures.

We have learned enough about these great earthquakes, previous prehistoric earthquakes, and current rates of strain accumulation to make meaningful assessments of the future, including plausible effects of future tsunamis. This region has, we believe, a high likelihood of generating a great earthquake within the next few decades — probably within the lifetimes of children now living along its coastlines. The tsunami that follows the earthquake will probably devastate the coastal cities, towns and villages of this part of western Sumatra, as well as the offshore islands. Tens or hundreds of thousands of people will die in this event, and the damage suffered will have effects for decades following, unless actions to significantly reduce the scale of the disaster begin now and are sustained over the coming decades.

Most of what we know about this dangerous section of the Sunda megathrust has come to us from palaeoseismic and geodetic research. To study the old earthquakes, we have turned to the biological rather than geological record — a record kept by the large coral colonies that are common on the fringing reefs of the offshore islands west of the Sumatran mainland. Coral organisms cannot tolerate much exposure to the air. Thus, the colonies grow upwards from their base as far as the waterline (specifically, up to the lowest low-tide level in a given year), after which growth continues only sideways. Once it reaches the waterline, it can only grow outward, forming a pancake-like colony, or "microatoll". Microatolls growing in these kinds of locations manifest the cyclical changes in elevation related to the earthquake cycle. Earthquakes accompanied by the dropping of the crust cause the top of the microatoll to drop beneath the water level, allowing it to grow upwards for several years without restraint until it reaches the water level again. If the crust is rising, the microatoll will actually die back if the rise is enough to lift the top of the microatoll out of the water (Figure 1.4a).

With the help of underwater chainsaws, we can take slab-like cross-sections of the microatolls. In these cross-sections, we can see annual growth rings, analogous to the growth rings of trees. By counting these rings, as well as by applying a radiometric dating technique, we can reconstruct the entire history of a microatoll's growth, which may extend back for well over a century. Furthermore, dead microatolls that record even earlier histories can be found. From these histories, we can deduce the dates of earthquakes reaching back for several centuries. By putting together the histories obtained from microatolls at many different locations, we can often reconstruct the extent and nature of the ruptures that caused the individual earthquakes, and thus obtain an estimate of their magnitudes.

From our analysis of the 1797 and 1833 events, we see that the earthquakes resulted from rupture of adjacent, slightly overlapping sectors of the megathrust, southeast of the rupture that caused the March 2005 earthquake (Figure 1.5).

Coral microatolls on the islands above the ruptures help us constrain the extent and magnitude of the two events.[5] The southern extent of the 1833 rupture is poorly constrained, but the size of the earthquake was probably between 8.7 and 8.9. A repetition of rupture of these sections of the megathrust now threatens about a million inhabitants of western coastal Sumatra. The Sumatran fault, which runs through the highlands of Sumatra and through Banda Aceh, the devastated capital of Aceh province, also poses a risk to Sumatrans.[6]

FUTURE SUMATRAN MEGATHRUST EARTHQUAKES AND TSUNAMIS

Starting in 2002, we added geodesy to our bag of scientific tools by beginning the installation of a network of continuously monitored GPS stations in Sumatra. We have set up twenty-seven of these stations so far, most of them

FIGURE 1.4A

Certain species of massive coral record changes in sea level, because they cannot grow above the sea surface. In this idealized cross-section through a coral colony, annual growth rings show it has grown up to sea level in five years. At year seven, it rose during an earthquake so the top of the coral was exposed above the sea and died. In the subsequent five years, the coral continued to grow outwards below the new sea level.

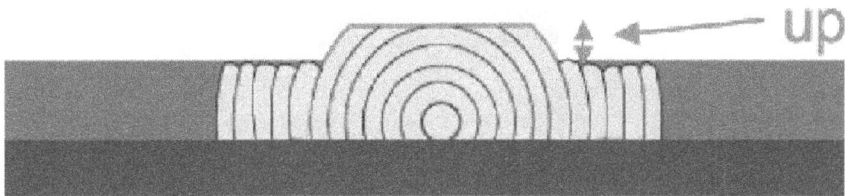

FIGURE 1.4B

This coral on Simeulue Island was mostly submerged below the sea until a foreshock of the great 2004 earthquake occurred in 2002. Uplift of about 15 cm during that M 7.3 earthquake caused the central perimeter of the head (indicated by the double arrow) to die. Two years of growth of the portion still below the sea ensued, but the remainder of the head died after uplift

on the offshore islands, only 20 km or so above the megathrust. There are also a few that have been installed on the mainland (Figure 1.6). This network detects current motions of the Sumatran crust with high precision. Most of the stations transmit their data to us via satellites, so we can monitor the data daily. The GPS data allow us to follow the ongoing, slow deformations

FIGURE 1.5

Great megathrust ruptures occurred in 1797 and 1833 in central western Sumatra (beneath the coloured patches).

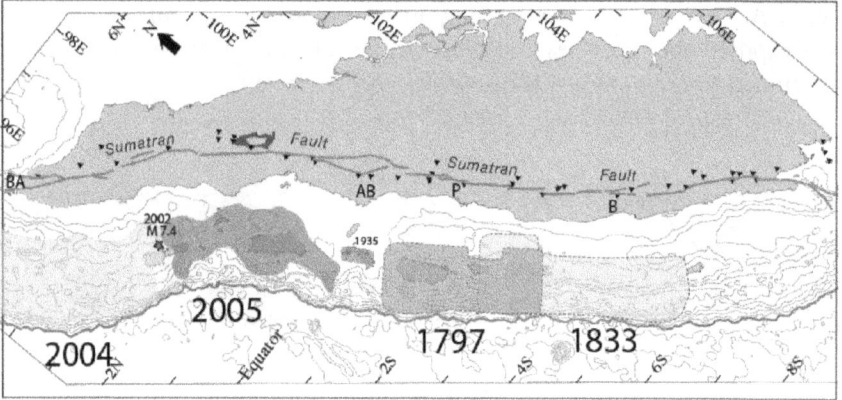

FIGURE 1.6

The Sumatran GPS Array currently consists of 27 continuously recording GPS stations.

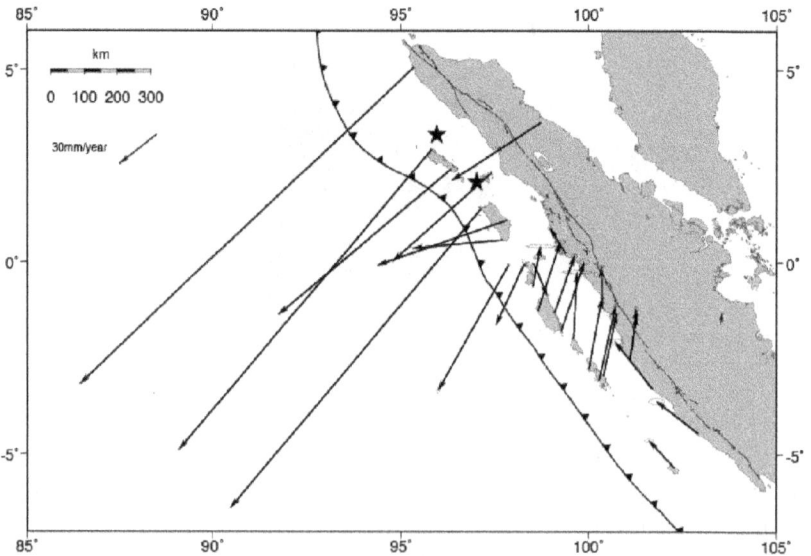

of the Earth's crust that go on in between earthquakes — in fact, they are better than the microatolls in that they record motions in both the vertical *and* horizontal directions. In addition, they detect the sudden displacements associated with the earthquakes themselves, such as the December 2004 and March 2005 events.[7]

Stations from the Equator north continue to show rapid adjustments to the great earthquakes of 2004 and 2005; large vectors pointing southwestward reflect after-slip on the megathrust throughout the year after the 2005 earthquake.[8] Stations south of the Equator show ongoing accumulations of strain that will be suddenly released during future great earthquakes there. Vectors from these stations show that the Mentawai islands and southern mainland coast are still squeezing because of the locking of the underlying megathrust. Corals show that this has been going on since at least the mid-twentieth century, and in all likelihood strains have been accumulating since the great earthquake of 1833.

North of the Equator

There is no historical record of an event comparable to the 2004 Aceh-Andaman earthquake. This is hardly surprising because, at the rate of steady plate convergence, it would have taken hundreds of years to accumulate enough strain to be relieved by the tens of metres of slip that occurred in 2004. Archaeological evidence on the east coast of India suggests, in fact, that the previous great tsunami occurred about a thousand years ago.[9]

In contrast, the 2005 earthquake appears to have an historical precedent 140 years earlier, in 1861.[10] This 140-year interval is almost precisely the interval one would expect, given the average amount of slip that occurred in 2005 (6 m) and the rate of convergence of the plates there (45 mm/year). Even though this interval in between quakes is far shorter than the many hundreds of years between 2004-like earthquakes, it seems that a repeat of the great 2005 earthquake is very unlikely within the next hundred years.

There is another type of earthquake, however, that may pose a risk to coastal residents of Nias and Simeulue. In 1907, an earthquake of modest magnitude (7.6),[11] produced a tsunami on the west coasts of those islands that was far higher than the tsunamis of 2004 or 2005. In fact, it was recollection of this tsunami that motivated people on Simeulue and Nias islands to flee to the hills after the 2004 and 2005 earthquakes — an action that ensured their survival. The source of the infamous 1907 tsunami is debated. Two sources, both west of the islands, are plausible: sudden rupture of the shallowest part of the megathrust, which has been creeping rapidly since the 2005 earthquake,[12] or rupture of a fault west of the Sunda trench on the oceanic seafloor. Recent mapping of bathymetry by our German colleagues shows that the oceanic seafloor is broken by numerous normal faults as it bends in preparation for descent into the subduction zone.[13] Both of these sources must be regarded as potential sources of a locally damaging tsunami in the next century.

South of the Equator

The question now arises whether the sector of the megathrust south of the Equator has been squeezed enough since 1797 and 1833 for it to rupture again in the near future. Evidently, it was not yet at the tipping point in 2004 or 2005, for if it had been, the March 2005 rupture would not have stopped where it did — it would have carried on southwards, past the Equator.

We can gain a better idea of where this sector lies in its cycle by examining its history over the past several earthquake cycles, as revealed in the coral records. These records show that uplifts as large as those in 1797 and 1833 also occurred in the late fourteenth and seventeenth centuries. Thus, it appears that great earthquakes (or earthquake couplets, as in 1797 and 1833) occur about every two centuries. This implies that we are in the last years or decades of the current dormant period and that another great earthquake is likely to occur in the near future, where "near" means not necessarily weeks or months or even years, but a few decades. The great earthquake could happen tomorrow or thirty years from now, but it is not likely to be delayed much beyond the next few decades.

Estimating Tsunamis South of the Equator

What would happen if the section of the megathrust south of the Equator were to rupture suddenly? First, significant damage and loss of life would likely be caused by the earthquake itself, particularly because many local buildings are inadequate to withstand the several minutes of strong shaking that would occur. But what about tsunamis?

In 1797, Padang was a tiny English colonial settlement one to two kilometres upstream from the coast on the banks of a small river. The tsunami ran up the river and, according to contemporary accounts, it seized a 150-tonne English sailing vessel that was moored near the river mouth, carried it up the river and deposited it over the river bank in the middle of town. That would have required an overland flow depth of several metres.[14] Padang is now a city of about 800,000 people that occupies nearly all of the first few kilometres from the coastline (Figure 1.7a) — clearly, the effects of a 1797-sized tsunami today would be horrific. The 1833 earthquake was likewise followed by a destructive tsunami, but it had less effect at Padang, which lay beyond the very northern end of the rupture zone. However, it did destroy the waterfront at Bengkulu, about 400 km to the south. Then a tiny settlement, Bengkulu now has a population of about 300,000. Altogether, there are more than a million individuals exposed to future megathrust earthquakes and tsunamis in Bengkulu, Padang and the other coastal cities, towns and villages of western Sumatra.

To date, the modest yet laudable post-2004 efforts by local NGOs and local government to mitigate tsunami hazard in Padang and the surrounding region[15] have relied on simple assumptions about future tsunamis. They have assumed that land lower than 5 m above sea level is dangerous, land between 5 and 10 m above sea level is relatively safe, and land above 10 m is safe. More precise information from scientists is sorely needed to aid in mitigation efforts.

To assess the specific effects of future tsunamis along this part of the coast, one needs to start with a plausible set of sources, that is, reasonable ruptures on the megathrust that produce uplift of the seafloor. Then, one must calculate the effects of the uplift of the seafloor on the sea itself. We have made an initial attempt at this.[16] First, we calculated the on-land effects of the 1797 and 1833 tsunamis, using the slips implied by the coral uplift during those earthquakes. The results were comparable to the sparse historical record of tsunami inundation and overland flow depths during those events. Armed with this success, we then calculated the effects for two plausible future scenarios: In one case, the entire 700-km length of the megathrust breaks, with 10 m of slip — an amount similar to that in the 2005 Nias-Simeulue earthquake, north of the Equator. In the other case, the slip is 20 m along this 700-km length — an amount similar to that in the 2004 Aceh-Andaman earthquake (Figure 1.8). The latter scenario can be considered a plausible worst-case scenario. In that case, a 700-km long welt, several metres high, develops on the sea surface west of the Mentawai Islands (Figure 1.8f). That welt spreads southwestwards into the Indian Ocean and northeastwards to the Sumatran coast.

The results of the models imply that much of the Sumatran coast south of the Equator would suffer destructive waves. At Padang and Bengkulu, the first cresting waves would strike about half an hour after the start of the earthquake (Figure 1.9). Bengkulu, unprotected by large offshore islands, experiences one long-lasting initial cresting wave, several metres high. By contrast, Padang experiences a set of three slightly smaller waves in the same time period, because the sea has to pass through the straits between large offshore islands on its journey to the city. Note that large waves also hit more than two hours after the earthquake. These are "edge" waves that move slowly along the coast in shallow waters.[17]

FIGURE 1.7A

Padang is now a sprawling city of about 800,000 people. Most of the town is less than 10 m above sea level.

FIGURE 1.7B

Many smaller towns and villages along the western coast of Sumatra will also be inundated by future tsunamis. The town of Air Bangis, near the Equator, has begun to prepare for the possibility.

FIGURE 1.8

Uplift of the seafloor produced by the six megathrust ruptures used in the study by Borrero et al. (2006).[1] (*A* and *B*) Dimensions of the 1797 (*A*) and 1833 (*B*) ruptures from Natawidjaja et al. (2006).[2] (*C*) In scenario 1, uniform slip of 10 m extends to trench. (*D*) In scenario 2, uniform slip of 10 m extends up-dip only to a depth of 15 km. (*E*) In scenario 3, uniform slip of 20 m extends to trench. (*F*) In scenario 4, uniform slip extends up-dip only to a depth of 15 km, and the seafloor bulges up about 8 m southwest of the Mentawai Islands.

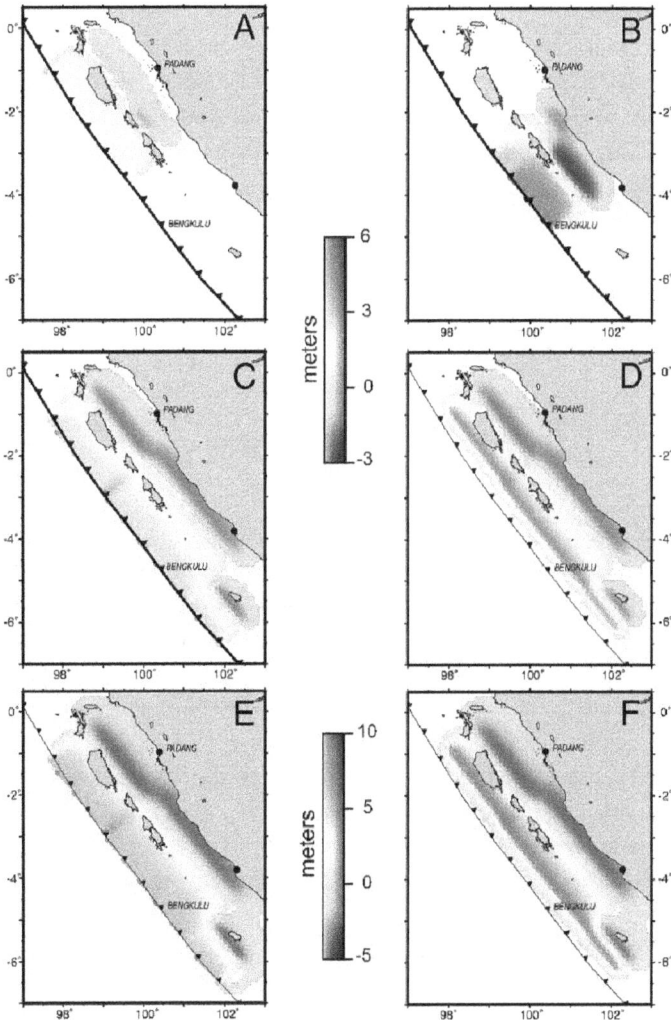

These simulations are just the beginning of what scientists must produce in order to aid mitigation of the tsunami hazard along western Sumatran shores. Refinement of the tsunami source will be realized through ongoing studies of the deformations now being recorded by the SuGAr array of continuously recording GPS instruments. Already, the new GPS data are suggesting that the sources used by Borrero et al.[18] are too large. Chlieh et al.[19] use the GPS data and the coral record of the past 50 years to show that the locked patch of the megathrust offshore from Bengkulu and Padang is no more than 600 km long, not the 700 km assumed in the models of Figure 1.8. This implies that future west Sumatran tsunamis will be somewhat smaller than in those models.

The topography and shallow bathymetry used in tsunami modelling also play critical roles in determining the local characteristics of tsunamis. The models in Figure 1.9 use the most detailed topography and bathymetry that are readily available — bathymetry from standard hydrographic charts and topography from NASA's Shuttle Topography Radar Mission (SRTM). Higher-resolution data would enable significant improvements in estimating future inundation distances and overland flow depths. Our German colleagues have announced plans to collect such data for use in a next generation of tsunami maps. This effort promises to provide still more precise and reliable data from which to construct tsunami hazard maps.

PUBLIC EDUCATION

It is beyond the scope of this chapter to discuss in detail the aspects of earthquake and tsunami hazard mitigation that should follow on the heels of the basic scientific definition of the problem summarized above. I have, however, explained my views about this in a previous paper.[20] Still, let me mention briefly what our group has done in the way of public education in Sumatra. In a nutshell, we have tried to teach people there why earthquakes and tsunamis occur. We also try to keep our work in public view, to encourage preparations, preparedness and change.

In the course of many visits, we have developed friendships with and admiration for many people in Sumatra. This has led us to proactively create an awareness of the hazards to which they are exposed. Starting in 2004, my colleagues and I began a programme of public education in the Mentawai Islands, which is where much of our research has been focused. The programme has had several elements. One is a set of posters that we distributed and put up in public spaces such as offices and businesses. These posters are in three languages: English (for the tourists and surfers), Mentawai

FIGURE 1.9

Maps of computed tsunami flow depths and inundations over local coastal topography near Padang (Upper) and Bengkulu (Lower) for model scenarios 1 and 3. Pixel dimension is 200×200 m. Below each map is the corresponding tsunami time series. Each begins at the time of fault rupture and is for offshore locations at water depths of 5 m (red dot) and 10 m (black dot). The solid black line represents the extent of densely populated urban areas. The simulations of 1797 and 1833 tsunamis are consistent with sparse historical accounts. Simulations of plausible future events (Scenarios 1 and 3) show that both Padang and Bengkulu could be seriously impacted by future tsunamis. The effects at Bengkulu, unprotected by offshore islands, will likely be more severe than at Padang.

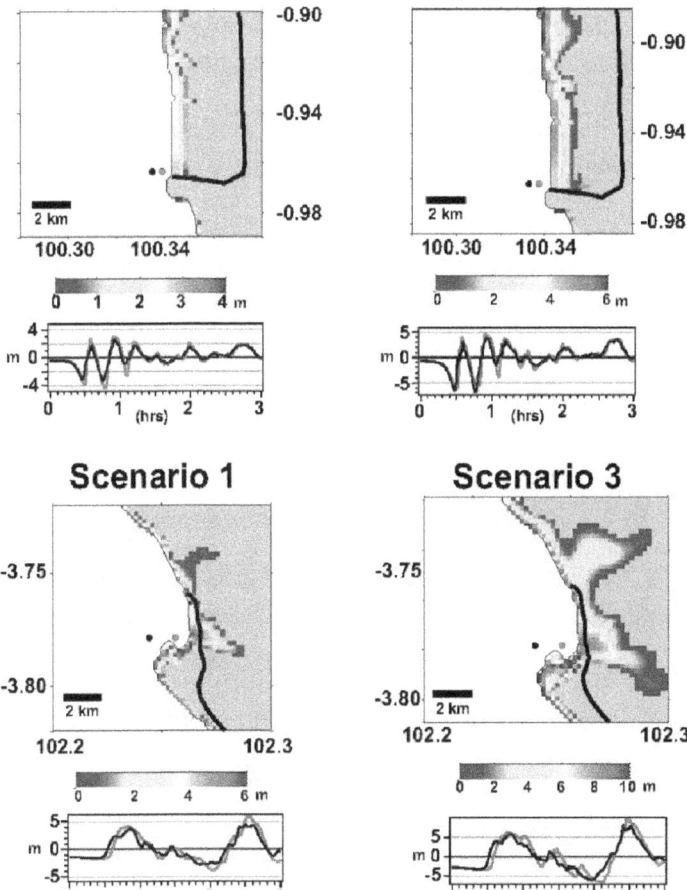

Scenario 1   Scenario 3

(the local language) and Indonesian. They explain our research and findings in a straightforward way and make clear what these findings mean in terms of earthquake and tsunami hazards. A small part of the posters introduce some of the steps that can be taken to reduce these hazards. Literacy rates are high, so these posters reach much of the population in the villages that we visit.[21] Figure 1.10 is our most recent poster, aimed at the populations on the mainland coast of Sumatra.

The 2004 and 2005 disasters have greatly aided our educational efforts. Sadly, it has taken disasters of this magnitude to draw public attention to earthquake and tsunami hazards, and the connection between them. The population of Padang was put in a state of great anxiety by the events to the north. The sequence of aftershocks mentioned earlier, which were situated close to Padang, caused great distress there, but of course this state of anxiety will gradually abate.

It is critically important to seize the opportunity brought about by the 2004 and 2005 disasters to educate local communities about the actions they can take to protect themselves from future giant Sumatran earthquakes. The main message that needs to be communicated is that people should respond to a long-lasting earthquake — say, one lasting 45 seconds or more — by running or cycling to high ground or inland, but that this is not necessary with the more frequent small earthquakes that typically last 10 to 15 seconds. In addition, people should be urged to support programmes for changes in infrastructure and development of emergency response capabilities.

THE INFRASTRUCTURE — BUILDING AND PLANNING FOR SAFETY

Another critically important aspect of hazard mitigation in western coastal Sumatra is dealing effectively with the enormous exposure of communities to shaking and tsunami inundation. The current situation is akin to an overnight campout in the middle of a not-so-busy street. Most of the night, the campers slumber peacefully. When a pair of headlights appears up the street, someone is supposed to wake everyone up so that they can flee before the car ploughs through the tents. Far better if the tents had been pitched in the front yard rather than in the street! Thus, the two key questions concerning how to build safely in coastal regions exposed to megathrust earthquakes and tsunamis are: first, where to build, and second, how to build.

As to the where, it is all too apparent that many Sumatrans now live in what at the time of a great tsunami are the wrong places — low-lying areas close to the ocean, estuaries or rivers. In the December 2004 tsunami and

FIGURE 1.10

The English version of the latest in our series of educational posters is aimed at coastal residents of mainland Sumatra. These posters present the science behind the earthquake and tsunami hazard. Additional efforts will be necessary to mitigate the hazards.

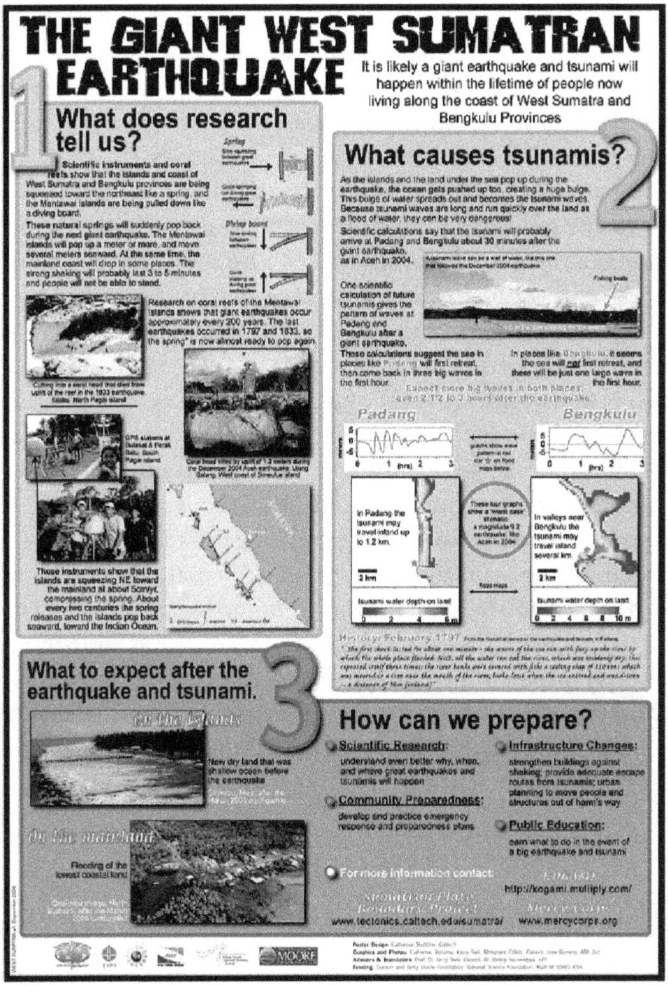

also in the March 2005 tsunami, low-lying areas in Aceh and North Sumatra were utterly devastated. In some coastal towns on the northwest coast of Aceh, barely a trace of human habitation remained after the tsunami; in such towns, 90 per cent or more of the inhabitants died. It will take decades or longer to recover from these immense losses.

The damage caused by these events was not limited to the destruction during the tsunami itself. In addition, the drop in the land surface that accompanied the rupture has left many coastal areas permanently underwater, as can be seen in satellite images of Aceh and North Sumatra provinces after the 2004 and 2005 earthquakes. These changes in the coastline will lead, over years or decades, to further destructive effects. The sea continues to eat away at the subsided coastal plain, moving the coastline even farther inland. Rivers, finding themselves flowing too steeply down to the ocean, will respond by flooding more widely and by cutting new channels.

All these processes, though completely natural and inevitable, will be very disruptive to the populations that are attempting to rebuild in the affected areas. How far away from the beach must a new coastal road be built? Where should bridges be located? How close to a riverbank can homes be safely built? To answer questions of this kind requires the expertise of coastal and fluvial geomorphologists, combined with a detailed understanding of local conditions. Nations with scientific and engineering capabilities in these areas could provide such experts, who would assist in the development of land-use plans and train Indonesian scientists in these fields.

As to the question of how to build, the issue is not tsunami resistance; rather, it is seismic resistance. Currently, building techniques on the islands, and to some extent on the mainland coast, are so elementary that quite inexpensive steps can be taken to strengthen people's homes against earthquakes. Most notably, a typical island home in places like Mentawai is supported by posts that are simply perched on blocks of coral laid on the ground. In even a moderate earthquake, such structures come off their "foundations" and, very often, collapse. This failure mode can be prevented by anchoring the posts to inexpensive concrete footings set 18 inches into the ground.

CONCLUSIONS

Scientific understanding of earthquakes and tsunamis provide the foundation for mitigating the effects of future earthquakes and tsunamis in Sumatra. The discovery of the potential for great earthquakes and tsunamis along the coast of Sumatra south of the Equator has laid the foundation for efforts at mitigation. Although more scientific work is needed, we have already identified, to first order, which megathrust patches off the shores of Sumatra are currently locked and therefore storing strain. And we have begun to calculate the characteristics of plausible tsunamis generated by future failure of these patches.

Most of the large, active megathrusts throughout South and Southeast Asia are not as well known as the one off the shores of western Sumatra. It should come as no surprise then that the hazards posed by these megathrusts are too poorly known to give much insight into the levels and specifics of preparation, education and change required. The succeeding paragraphs provide a few glaring examples.

The danger posed by the section of the Sunda megathrust off the southern shore of Java, one of the most densely populated coasts on Earth, is not known. Based on the tsunami disaster of July 2006, it seems that even moderate earthquakes there have the potential to generate locally dangerous tsunamis. But is it possible that the megathrust south of Java could also generate a much larger earthquake, say an 8.5 or 9, which would produce a far more devastating tsunami along one of the most densely populated coasts on Earth? A programme of continuous geodetic monitoring would be one simple step towards answering this question.

As to which sections of the Himalayan megathrust will probably break next, it is well known that this giant fault produces great earthquakes from Pakistan through India and Nepal to Bangladesh.[22] However, next to nothing is being done to address the vulnerabilities of the populations it threatens. One can hope that a greater level of specific scientific information would motivate serious mitigation activities. Lacking such efforts, one can only wonder which megacity of the Gangetic plain will be the first to lose the gamble that it has placed by virtue of its inattention to the problem.

What is the potential of the Manila megathrust, which traverses 1,200 km of the South China Sea from the Philippines to Taiwan? If its entire length were to rupture in one event, the earthquake and ensuing tsunami would rival the Indian Ocean event of 2004. Yet no one knows whether this is possible. Palaeo-tsunami studies of coastal Vietnam and southern China could begin to address this question. Geodetic investigations would be more difficult, because there are no islands near the megathrust upon which to site GPS stations. Nonetheless, the prospect is important enough to communities on the coast of the South China Sea (such as Macau and Hong Kong) that an expensive submarine GPS system might be warranted.

Throughout South and Southeast Asia, scientists do not have the public and governmental support necessary to effectively evaluate the potential for large earthquakes and tsunamis. As a result, tragedies like the 2004 tsunami come as great surprises. Moreover, the other important links in the hazard mitigation chain — public education, emergency response preparedness and infrastructural resilience — do not occupy a prominent place on most agendas. Maintenance of this status quo throughout South and Southeast Asia will prove tragic and expensive, for without strong chains linking scientific discovery to mitigation, other events as profoundly disturbing to human well-being as the 2004 tsunami will strike elsewhere in the coming century.

Even along the western coast of Sumatra, where scientific studies are well advanced, the links between science and mitigation efforts are weak. For although scientific discovery there has shown that a large earthquake and tsunami are very likely within the next few decades, activities aimed at reducing the exposure of the coastal communities are too meagre to make more than slight progress in solving the problem.

One test of whether humanity acts responsibly in the next millennium is this: Can we marshal the visionary persistence needed to take charge of our future? Or will we carry on as we did throughout most of the past — simply reacting to tragedies as they happen? If the answer is the latter, then there will continue to be more tragedies like that of 26 December 2004.

Notes

1. K. Sieh, "Sumatran Megathrust Earthquakes: From Science to Saving Lives", *Philosophical Transactions of the Royal Society* 364 (2006): 1947–63.
2. C. Subarya et al., "Plate-boundary Deformation Associated with the Great Sumatra-Andaman Earthquake", *Nature* 440 (2006): 46–51; M. Chlieh, J. Avouac, K. Sieh and D. Natawidjaja, "Investigation of Interseismic Strain Accumulation along the Sunda Megathrust, Offshore Sumatra", *Journal of Geophysical Research* (in review).
3. R. Briggs et al., "Deformation and Slip along the Sunda Megathrust in the Great 2005 Nias-Simeulue Earthquake", *Science* 311 (2006): 1897–1901.
4. D. Natawidjaja et al., "Source Parameters of the Great Sumatran Megathrust Earthquakes of 1797 and 1833 Inferred from Coral Microatolls", *Journal of Geophysical Research* 111 (2006), doi:10.1029/2005JB004025.
5. Natawidjaja et al., "Source Parameters".
6. K. Sieh and D. Natawidjaja, "Neotectonics of the Sumatran Fault, Indonesia", *Journal of Geophysical Research* 105 (2000): 28, 295–28, 326; S. Nalbant et al., "Earthquake Risk on the Sunda Trench", *Nature* 435 (2005): 756–57.
7. Briggs et al., "Deformation and Slip".
8. Y. Hsu et al., "Frictional Afterslip Following the 2005 Nias-Simeulue Earthquake, Sumatra", *Science* 312 (2006): 1921–26.
9. C. Rajendran et al., "The Style of Crustal Deformation and Seismic History Associated with the 2004 Indian Ocean Earthquake: A Perspective from the Andaman-Nicobar Islands", *Bulletin of the Seismological Society of America* (in press).
10. K. Newcomb and W. McCann, "Seismic History and Seismotectonics of the Sunda Arc", *Journal of Geophysical Research* 92 (1987): 421–39.

11. B. Gutenberg and C. Richter, *Seismicity of the Earth and Associated Phenomena* (Princeton, NJ: Princeton University Press, 1954).
12. Y. Hsu et al., "Frictional Afterslip"; F. Tilmann et al., "First Results from a Combined Marine and Land Passive Seismic Network near Simeulue Island", *Eos, Transactions of the AGU, Fall Meeting Supplement* 87.52 (2006): U53A-0025.
13. M. Schauer et al.,"Morphotectonics of the Sumatra Margin — Analysis of New Swath Bathymetry", *Eos, Transactions of the AGU, Fall Meeting Supplement* 87.52 (2006): U53A-0033.
14. Natawidjaja et al., "Source Parameters".
15. <http://multiply.com/i/xqeWVamMRpEvgXAiTyIvjw>.
16. J. Borrero, K. Sieh, M. Chlieh and C. Synolakis, "Tsunami Inundation Modeling for Western Sumatra", *Proceedings of the National Academy of Sciences* 103 (2006): 19673–77.
17. Movies of the simulated tsunamis can be downloaded from <http:// www.pnas.org/ cgi/content/full/060469103/DC1> and <www.tectonics.caltech.edu/sumatra/ tsunami_models.html>.
18. Borrero et al., "Tsunami Inundation".
19. Chlieh et al., "Investigation of Interseismic Strain".
20. Sieh, "Sumatran Megarthrust Earthquakes".
21. Copies of the posters are available for downloading at <www.tectonics.caltech. edu/ sumatra/public.html>.
22. R. Bilham, "Dangerous Tectonics, Fragile Buildings, and Tough Decisions", *Science* 311 (2006): 1873–75.

PART I
Reconstruction Efforts

2

DISASTER RECOVERY
An International Humanitarian Challenge?

John Telford

DISASTER RECOVERY: DEFINITIONS, DESCRIPTIONS AND CONSTRAINTS

Natural disasters produce long-term and complex impacts on survivors' livelihoods, on their physical, social and political infrastructure, and the environment. In almost all cases, recovery operations appear almost immediately following a natural disaster. After being initially absorbed with helping survivors locate loved ones and organizing emergency aid, recovery efforts rapidly turn to longer-term concerns such as housing, re-opening schools and re-establishing income generation and livelihood activities. As is argued by other authors in this volume, to be successful, recovery activities must be rooted firmly in local and national priorities, processes and capacities. This, however, has become increasingly complicated because of the large-scale internationalization of disaster response efforts.

In industrialized countries, natural disaster response is typically managed ("owned") by the affected states, as can be seen in the US response to Katrina, the Japanese response to the Kobe earthquake, and the more recent Chinese response to the Sichuan earthquake. Disaster-affected persons are the primary actors in their own recovery, funded and led in large part by national and regional authorities. Respect for such "ownership" is a principle of international humanitarian aid, as reflected in the Sphere Project standards, the Red Cross Code of Conduct and the Good Humanitarian Donorship initiative.[1]

In "non-industrial" nations, and other instances where the scope of the disaster exceeds local capacities to respond effectively, a wide range of humanitarian actors have increasingly become involved. This is clearly illustrated by the intensive response to the 2004 Indian Ocean Tsunami, especially in Indonesia and Sri Lanka. A dramatic increase in the funding entrusted to international humanitarian aid actors has put pressure on them to assume more prominent roles in post-disaster settings (see section below on funding). However, they face major constraints and challenges in designing and implementing recovery programmes. While there are many cases of "good practice", in this chapter I will be focusing more on the appropriateness and effectiveness of humanitarian aid actors in the face of such recovery challenges and constraints. In doing so, I concentrate more on the shortcomings of international organizations.[2]

This chapter is based on a number of sources. These include primarily the Tsunami Evaluation Coalition (TEC) Synthesis and other TEC reports, which examined the international response to the tsunami. These reports focused on Indonesia and Sri Lanka, while also examining the response in the Maldives and Thailand. As one of the authors of the TEC study, I make use of some of our findings to explore some of the shortcomings of the humanitarian response to the tsunami, particularly in Aceh. To more fully contextualize the tsunami response, this chapter also draws on the author's previous experiences with recovery processes following disasters in Honduras (1998), Colombia (1999), El Salvador (2001), India (2001) and Iran (2003). I start by going over some of the key points that the TEC flagged, and then move on to discuss a number of key points in greater detail.

International agencies are criticized in TEC reports for making programming choices without adequately consulting and considering affected people and the contexts in which they live. This is in part the result of the relative power and wealth of agencies, which, the reports claim, overwhelmed and undermined national and local capacities: "local ownership ... was undermined and some local capacities were rendered more vulnerable" and "treating affected countries as "failed states" was a common error".[3] Even where local and national capacities were recognized, they were often exploited to strengthen international organizations rather than to underpin local responses.

We found that international organizations frequently failed in the modest objective of informing affected people in an accurate, timely and comprehensive manner: "A tragic combination of arrogance and ignorance has characterized how much of the aid community ... misled people ..."[4] False promises to affected people were an especially galling example of international arrogance. False *partnerships* (i.e., arrangements which amounted to little more than convenient outsourcing to local organizations) are another example. Other problems identified in the TEC thematic evaluations and their sub-studies include brushing aside or misleading authorities, communities and local organizations; inadequate support to host families; displacement of able local staff by poorly prepared internationals; the application of more demanding conditions to national and local "partners" than to international agencies; "poaching" staff from national and local entities; "misrecognition" of local capacities, resulting in inefficient implementation; and limited participation of the affected population. Such practices led to inequities, gender- and conflict-insensitive programming, indignities, cultural offence and waste.

I have come across similar practices in other disasters. They can be traced back to the following root causes: the paucity of international and national legislation governing the obligations of international agencies in disaster-affected countries; the poor capacity, preparation and tools of many international humanitarian staff (regarding, for instance, communication skills, contextual knowledge and sensitivity, and tools for identifying local capacities); and the monopoly held by certain international agencies as conduits for international humanitarian aid. In the next section of this chapter, I move on to talk about the overall organization of the aid efforts, and a number of issues related to donors and funding. These are critical areas that need to be re-addressed by the international community to provide more inclusive, responsive and effective humanitarian aid responses to future human and natural catastrophes.

DISASTER RECOVERY IN INDONESIA: RELIEF AGENCIES AND THE BRR IN ACEH AND NIAS, 2005[5]

The first step is to look in more detail at the official mechanisms for disaster response in Indonesia before the tsunami, their shortcomings, and the consequences of altering administrative frameworks on the fly. Indonesia has developed relatively strong national institutions and legal systems, including the Disaster Management Coordination Board (BAKORNAS) and the National Development Planning Board (BAPPENAS). However, disaster management capacities, especially at the regional and local levels, had not been designed for catastrophes like the tsunami. Gaps, weaknesses and the poor definition of roles and responsibilities became particularly evident in the early recovery phase. BAPPENAS was a central actor in the design of the reconstruction master plan (or "blueprint"). The plan included the establishment of a new Rehabilitation and Reconstruction Agency (*Badan Rehabilitasi dan Rekonstruksi Aceh dan Nias*, BRR). The BRR was launched in April/May of 2005, but did not become fully operational until later in the year. It was tasked with coordinating and overseeing the reconstruction effort under the direct authority of the president of Indonesia. One of its main roles was to provide international organizations with an official Indonesian authority with whom they could work.

That special role brought with it complications. Chronic armed conflict in Aceh and the predominant role of the Indonesian military had weakened local institutions in Aceh over the years, and greatly reduced Acehnese trust in officials from Jakarta. The tsunami compounded this process through its massive human and material impact and by further weakening local physical and social infrastructure. One of the major issues that came to the surface right away, and which continued to be a problem for the life of the BRR, was the perception that they were an outside institution. The BRR was widely seen on the ground as an Indonesian government body, which has implications for its reception by many Acehnese who had a hostile relationship with the federal government. Regardless, all of the international humanitarian organizations were required to work with the BRR.

The establishment of the BRR was seen by many international agencies as an important step for giving direction and transparency to aid efforts. Though criticized, recognition by international actors of its focus, authority and gradual improvement of systems (such as information management and sectoral, intersectoral and cross-cutting expertise) grew throughout 2005 and 2006:

> After a period of uncertainty about the required long-term presence of international agencies in Aceh, the creation of the Aceh and Nias Rehabilitation and Reconstruction Agency (BRR) in April 2005 introduced greater confidence and closer coordination between all stakeholders.[6]

Compared with the massive funds available and the huge number of organizations present, international financial and technical assistance to the local authorities for coordination was relatively minor and late. Even cooperation was sometimes not forthcoming.[7] Support such as that from the UN Information Management System (UNIMS) to the BRR, ministries and district and subdistrict administrations was more the exception than the rule. Local authorities that had responded in initial relief operations found themselves relegated to a secondary role during the early recovery phase.

Despite the establishment of the BRR as a single planning and coordination body, a multiplicity of coordination mechanisms existed, particularly those involving donors. In 2005, the BRR organized coordination meetings roughly every two weeks in Jakarta, involving donor representatives. It also proposed a Coordination Forum for Aceh and Nias (CFAN) "with the explicit aim of engaging a wider range of stakeholders — bilateral and multilateral donors, international non-governmental organizations (INGOs), civil society and central and local government".[8] Additionally, the $500,000,000 USD multi-donor trust fund (MDTF) established a governance and coordination mechanism for fourteen main members. However, the MDTF represented only a few of all donors and only a part of all funds. In fact, much of the funding of MDTF members was managed outside the MDTF. Furthermore, a plethora of other, often overlapping coordination mechanisms existed, based on the consortia, agency alliances and "families", and formal and informal networks that make up the complex web of modern international aid.

Poor coordination capacities led to striking imbalances in certain locations between aid agency presence or activities and the real needs of the Acehnese. In mid-2005, areas with relatively heavy concentrations of agencies and funds included Banda Aceh and Aceh Besar, to the detriment of neighbouring districts such as Aceh Jaya and Pidie, or further afield in the south and northeast of Aceh and in Nias. Only in 2006 did the BRR begin "to address the problem through the promotion of interagency policy advisory groups and the creation of subdistrict coordination forums".[9]

Furthermore, as part of its planning and coordination function, the BRR was established to play a focal role regarding quality control, accountability and transparency. It developed, perhaps somewhat late during 2005, forty monitoring and evaluation indicators. The BRR also provided aid agencies and beneficiaries with housing guidelines, covering construction design, total area and materials. To improve transparency and accountability, especially in cases where confusion or misunderstanding arose between beneficiaries and aid agencies, the BRR supported mass information campaigns, including the use of posters, village billboards and radio dissemination. Secondly, it appointed an ombudsman to deal with specific cases. A third measure was the establishment of an Anti-Corruption Unit (SAK) in September 2005. In two months, the unit examined over a hundred allegations.

Unfortunately, the BRR lacked the capacity, means and, perhaps, the will to formally control, regulate or certify NGOs. Some saw this as a function of the UN (specifically, the Office for the Coordination of Humanitarian Affairs, or OCHA), which was equally ill-designed and ill-prepared for such an approach. To better understand some of the issues with the BRR and the overall tsunami response, it is important to look in more detail at funding. Regardless of the mandate of the BRR, it is clear that a tremendous amount of pressure and influence was related to funding. In the next few sections, I discuss some of the main issues related to the procurement and use of funding that made the tsunami response unique compared to many other large-scale disaster aid efforts.

DISASTER RECOVERY: TSUNAMI FUNDING PROBLEMS

While previous disasters (such as Hurricane Mitch) also drew large funding, the tsunami was the most rapidly and generously funded disaster response in history: at least US$13.5 billion was pledged or donated internationally for emergency relief and reconstruction, including more than US$5.5 billion from the general public in developed countries. The total economic cost of the damage, and the consequent losses, were estimated at US$9.9 billion across the affected region, with Indonesia accounting for almost half of the total. Additional billions of US dollars in donations have gone unrecorded (such as remittances from abroad and national and local contributions). Private donations broke many records;[10] governments were flexible and quite rapid in their funding, and the reporting of pledges and commitments was better than in other crises.

Troubling aspects related to funding have emerged from the tsunami response. First, most private funding was concentrated in the hands of a dozen of the main actors. INGOs and the Red Cross/Red Crescent movements often had more funding than donor administrations or multilateral organizations. Few international organizations tried to halt fundraising when limits were reached. The TEC Needs Assessment Report (2006) sums up the impact of generous funding on implementing agencies as follows:

> Generous funding not only exceeded the absorption capacity of an overstretched humanitarian industry and deprived it of its customary excuse for built-in systemic shortcomings, but also led to the proliferation of new actors with insufficient experience (and therefore competence), as well as to established actors venturing into activities outside their normal area of expertise. Finally, the relative excess of funding was a disincentive to assess, to coordinate and to apply the results of the few collective assessments.

Second, similar to many other crises, governments and international organizations failed to ensure that funding was "needs based". Imbalances, non-needs-driven motivations (including funding INGOs based in a donor's own country, regardless of whether they had any comparative advantage over other NGOs in terms of addressing the needs on the ground), poor "end-user" traceability and inadequate monitoring were evident among official donor responses: "Allocation and programming, particularly in the first weeks and months, was driven by politics, funds and contextual opportunism, not by assessment and need."[11] The allocation of funds was fairly evenly split between relief and reconstruction. This did not reflect the reality that longer-term reconstruction and development needs are vital for sustainable and lasting recovery, and typically require far more resources spread over a much longer period of time.[12]

Competition among donors for visibility and perceived political interest drove up funding in a frenzy of counter-bidding, likened to a beauty contest by an EU Commissioner. In the case of Hurricane Mitch, important funding decisions had been motivated as much by political declarations and geopolitical "backyard politics" as by professional assessments of needs and capacities. Interestingly, government or intergovernmental funding was more for Mitch-affected countries than for tsunami-affected countries (US$9 billion pledged for the former versus some US$8 billion for the latter).

Third, official funding was not based on systematic measurements of the relative effectiveness and efficiency of organizations and their programmes. The reality that the capacity to deliver appropriate results did not match this largesse did not seem to matter. Few organizations were capable of putting the vast amounts of funding collected to effective use, especially under the time frames demanded by many donors. Additionally, the influx of money attracted hundreds of international agencies regardless of expertise and relevance, resulting in a highly fragmented response.

Fourth, weaknesses were evident regarding transparency on the use and destination of funds. The lack of system-wide definitions and standards for reporting funds was exacerbated by the cascading layers of contracts and subcontracts among international, national and local organizations, many of which did not deliver measurable "value added" as they passed on funds, minus considerable overhead charges. The local flow of financial information to affected populations in their own languages was especially weak.

Fifth, international agencies are increasingly becoming commercial organisations with a brand to project and protect. Their survival depends as much on how they capitalize on emergencies to generate income as on their actual performance. Large "communications" departments work assiduously to convince donors (both official and private) that they should receive money to help disaster-affected people. Appeals highlighting acute emergency conditions are more likely to raise money that those presenting more mundane, though possibly more important, reconstruction needs. This resulted in expectations that the money raised for the disaster response should be spent rapidly, often exceeding what was possible or appropriate. I discuss the implications of this in more detail below.

DISASTER RECOVERY: GLOBAL FUNDING ISSUES

My experiences dealing with a number of post-disaster situations show that the principal drivers for recovery are the affected people themselves, with support from their family or friends. In the case of Honduras,

> people's own skills, efforts, and resources, based on family remittances from abroad, family or individual labor and savings, and/or standard rate bank loans, were probably the main source of family support during the recovery period.[13]

The TEC reports also highlight the importance of remittances from members of diasporas, as support to national and local capacities. International funding is thus only a complement, albeit an important one, to national and local contributions, rather than the only source of contribution, as frequently portrayed.

Humanitarian aid is a booming business. It has grown steadily over the past two decades. Globally, annual humanitarian aid allocations from all recorded sources are estimated at some US$10 billion.[14] While it is difficult to argue that development aid funding has decreased in absolute terms, the relative significance of humanitarian aid has *increased*, despite international commitments to, for example, the Millennium Development Goals (MDGs).

Most humanitarian funds are donated for major emergencies, whether acute or chronic. This results in highly irregular funding flows: years of plenty interspersed with periods of much reduced funding. Additionally, funding levels vary immensely from one emergency to another. Some are massively funded (the tsunami response and, to a lesser degree, Hurricane Mitch), while others go relatively unnoticed by international donors and the public. This is especially the case if they occur far from donor borders; if they do not have important implications for donors' geopolitical interests; and if armed conflict is a major element of the crisis, as in the chronically underfunded Darfur and Democratic Republic of the Congo emergencies.

Often, international humanitarian aid is a supply-driven phenomenon, especially when dealing with non-monetary contributions. Typically, whatever is available or convenient is donated rather than only what is needed. In fact, it often seems that donors are largely unaware of how best to address the vital needs on the ground. The Gujarat earthquake was a case in point. While people requested assistance in their quest for permanent shelter and for access to income-generating opportunities, international agencies shipped masses of unsolicited goods which remained unused, even unopened, in many cases. Massive deliveries of inappropriate and secondhand clothing and medicines to disaster sites have become a caricature of bad donorship practices. The tsunami responses presented numerous similar cases, resulting in waste, confusion and even offence.

Special, "unearmarked"[15] funding, based more on people's desires to give than on proven need, gave agencies an exceptional level of independence, especially large INGOs. This provided an opportunity for flexible and appropriate recovery activities. More frequently, however, it led to perceived pressure to spend faster and more visibly than was advisable. The result has been fragmented, poorly coordinated, inappropriate and low-quality initiatives instead of the promised "build back better" approach.

DISASTER RECOVERY: STRUCTURAL PROBLEMS IN AID ADMINISTRATION

The "relief-development aid divide" is rooted in the administrative need to differentiate budgets. Given the inflexible and cumbersome procedures built up over decades around official development aid disbursement, alternative, more agile and simpler funding instruments became necessary for dealing with emergencies. Once created, they had to be differentiated in some way from development budgets, to ensure that non-emergency programmes did not benefit from the special facilities associated with emergency funding.

The most common differentiation is related to timing. Commonly, emergency funds must be spent within a relatively short period, often a matter of months. The erroneous equation of "emergency" with "humanitarian" has meant that non-acute emergency response expenditures (for example, for longer-term recovery and disaster risk reduction) are often time-limited in the same manner. Appeals by agencies for funds using dramatic images and language depicting emergency conditions add to the general expectation and, in some cases, legal requirement that humanitarian funds be spent rapidly, irrespective of how much time activities and programmes should actually take (in some cases years rather than months).

Official U.S. funds for reconstruction in post-Mitch Honduras had to be spent within a tight time frame[16] despite the fact that some activities could and should have been allowed more time for effective and efficient completion. The French Cour des Comptes public spending oversight body found that nearly half of the EUR323 million (US$428.8 million at the time) in aid gathered from public and private donors for tsunami-affected areas had not been used by 31 December 2005. It requested that the unspent balances be re-allocated to "other charitable causes".[17]

The frequent result is rushed implementation at considerably higher cost than might have been necessary. Roads and bridges in Honduras, houses in Bam and Gujarat, and income generation in tsunami-affected countries are all examples of this unfortunate and unnecessary trade-off among time, cost and quality. Some of the expensive, yet shoddy and inappropriate recovery efforts can be traced back to this administrative differentiation of funding budget lines.

A second drawback is that closely related activities (for example, provision of transitional shelters and longer-term housing reconstruction and income generation) are often funded from diverse budgets, and designed and managed under separate assumptions using separate data by separate staff, based in separate physical locations and under separate timelines. Humanitarian funded disaster recovery efforts are, as a result, rarely integrated with pre-existing or new development programmes and vice versa, thus reducing opportunities for Linking Relief, Rehabilitation and Development (LRRD).

DISASTER RECOVERY: CORRUPTION AND INTERNATIONAL AID

A recurring argument for the use of international agencies as conduits for funding (be it from private or public sources) is the avoidance of corruption. Corruption concerns have been raised in many post-disaster responses, and not just in the developing world. Reports of fraud were also common regarding disaster response in New Orleans. Allegations of massive corruption in economically developed countries are disturbingly common.[18]

It is clear that special vigilance and controls to ensure financial probity are necessary. What has not been proven, however, is whether international organizations can ensure transparent and efficient use of disaster funds any better than national entities. Apart from cases such as those referred to above, the excessive costs, duplication and waste associated with international disaster responses raise serious questions, and the assumption that international organizations are a *sine qua non* for ensuring financial transparency and accountability merits examination.

CONCLUSIONS AND RECOMMENDATIONS[19]

International organizations will continue to have a role in disaster recovery, given the all-too-frequent occurrence of large-scale human and natural catastrophes. The challenge is to redefine that role and adapt humanitarian capacities to best meet it. The structures, frameworks, capacities and motivations underpinning international humanitarian aid need to be reformed to ensure more efficient and appropriate approaches to post-disaster recovery. Below I list a number of areas that critically need to be addressed.

Ownership, Regulation and Accountability

The international humanitarian community needs fundamentally to reorient from supplying aid to facilitating communities' own recovery priorities. This necessitates a commitment to devolve control of resources and decision-making to affected people as much as possible. States should set standards and procedures for inviting, receiving and regulating international assistance. These should be codified by developing an International Disaster Response Law (IDRL) as a means of clarifying and strengthening the respective responsibilities, accountabilities and authority of affected states, international agencies and bilateral actors, such as international military forces. In addition, states and international agencies should establish an accreditation and certification system to distinguish agencies that work to a professional standard in particular sectors.

Concentrate on Disaster Risk Reduction

Disaster Risk Reduction (DRR) should be a declared priority for all states in high-risk regions and for the international agencies purporting to assist disaster-affected people. Comprehensive, multi-year risk reduction programmes should be anchored within national and local community development and social protection initiatives. Donor governments and international financial institutions (IFIs) should consider allocating a set percentage of their relief budgets to DRR; funding should be long term and predictable.

Geographic and Technical Specialization

Given that international agencies cannot be present in every country at risk, they should focus on a reduced number of high-risk regions or countries. This would allow the development of long-term partnerships and locally relevant knowledge and skills (including language capacities). In the same spirit, agencies should develop technical specializations in which they would be recognized as experts and held accountable for related performance and results. The discipline to say no to opportunistic deployments would thus prevail in favour of a more effective and rational international response system. Priority specializations for disaster recovery include the three areas of transitional and permanent shelter, livelihoods and disaster risk reduction.

Structural Change: A Flexible Aid Model

Humanitarian, transition and development aid (including multi-year programming) covering preparedness and prevention, emergency response and development programmes, should be administered by units and staff with the authority, knowledge and skills to provide aid as contexts require. If a fully integrated flexible model is not feasible, the staff managing development funding should at least become involved in recovery responses from the beginning of a disaster.

Reforming Fund-raising, Fund-giving and Fund-receiving

All actors need to make the current funding system impartial, more efficient, flexible (including multi-year funding), transparent and better aligned with principles of good donorship. International agencies should develop thresholds for their fundraising based on assessed need and their capacity to deliver effectively and efficiently. Any future accreditation/certification system should include conditions on how money is solicited and, as necessary, re-allocated to other actors or programmes (for example through trust-fund arrangements).

Donors should develop mechanisms for measuring the relative effectiveness, efficiency and accountability of international agencies, and fund accordingly. They should also inform the taxpaying public of the performance and quality of the agencies they fund, through, for example, widely disseminated independent reports. Donor, public and media education is necessary to improve understanding of and support for the above lessons and recommendations, including recognition of local contributions and diaspora remittances when reporting total funds donated for relief and recovery.

Summary

In this chapter, I have used some of the findings of the TEC as well as years of experience working in the humanitarian aid field to highlight what I see as some of the main issues that arose during the first several years of aid and relief efforts following the 2004 Indian Ocean Tsunami. A number of factors converged to change the lives of millions of people for the better, but many mistakes were also made that adversely affected people throughout the region. Given the scale of the response, and the vast amounts of resources spent, it is important to use this as an opportunity to rethink some of the operations of international actors in the aftermath of disaster. Hopefully the work of the TEC represented in this chapter contributes towards reforming the mechanisms of post-disaster humanitarian aid.

Notes

1. Humanitarian aid is defined as "Assistance, protection and advocacy actions undertaken on an impartial basis in response to human needs resulting from complex political emergencies and natural hazards". Disaster preparedness, prevention and recovery activities are all commonly included under the term humanitarian action, which now reaches well into non-emergency activities. The Good Humanitarian Donorship (GHD) initiative states that:

   > The objectives of humanitarian action are to save lives, alleviate suffering and maintain human dignity during and in the aftermath of man-made crises and natural disasters, as well as to prevent and strengthen preparedness for the occurrence of such situations (GHD 2003).

2. Of course there are significant problems that arise from local issues such as weak national and local capacities; armed conflicts; vacillating, ill-advised and corrupt national and regional leadership; politicized, bureaucratized and overcentralized decision-making; and complex issues related to land ownership, shortages, spatial planning and environmental considerations. All of these played key roles within the recovery processes in Aceh and Sri Lanka, and to a lesser degree in other tsunami-affected areas.

3. Elisabeth Scheper, Arjuna Parakrama and Smruti Patel, *Impact of the Tsunami Response on Local and National Capacities* (London: Tsunami Evaluation Coalition, 2006).

4. Ian Christoplos. *Links between Relief, Rehabilitation and Development in the Tsunami Response* (London: Tsunami Evaluation Coalition, 2006).

5. This section examines the relationship and interaction between international agencies and the official recovery agency in Indonesia following the 2004 tsunami. It draws principally on Tsunami Evaluation Coalition (TEC) "thematic reports", research and fieldwork which were conducted mainly in 2005. Claude de Ville

de Goyet and Lezlie C. Morinière, *The Role of Needs Assessment in the Tsunami Response* (London: Tsunami Evaluation Coalition, 2006).
6. J. Bennett, W. Bertrand, C. Harkin, S. Samarasinghe and H. Wickramatillake, *Coordination of International Humanitarian Assistance in Tsunami-Affected Countries* (London: Tsunami Evaluation Coalition, 2006), p. 40.
7. "For example, by October 2005 there were an estimated 438 NGOs working in Aceh, only 180 of which had reported on their activities to the government". See Christoplos, *Links between Relief, Rehabilitation and Development*, p. 37.
8. Bennett et al., *Coordination of International Humanitarian Assistance*, p. 65.
9. Ibid., p. 57.
10. The term "private" covers both the general public and private entities such as companies, religious groups or associations; in other words, all non-institutional donors. The bulk of these donations came from private individuals.
11. Michael Flint and Hugh Goyder, *Funding the Tsunami Response* (London: Tsunami Evaluation Coalition, 2006).
12. It should be recognized, however, that *some* donors strongly favoured recovery or reconstruction over emergency relief activities.
13. J. Telford, M. Arnold, A. Harth with ASONOG, "Learning Lessons from Disaster Recovery: The Case of Honduras", Disaster Risk Management Working Paper Series No. 8, Washington, DC: World Bank, 2004.
14. ALNAP, *Review of Humanitarian Action 2003* <http://www.alnap.org/publications/rha.htm> (accessed 12 May 2009).
15. Funds that are donated free of any conditions or restrictions set by the donor. They can be spent on virtually anything the recipient agency deems appropriate.
16. A U.S. Congress limitation stipulated that funds be spent by 31 December 2001.
17. Emmanuel Jarry, "French Watchdog Says Reallocate Excess Tsunami Aid", *Reuters*, 3 January 2007 <http://www.reliefweb.int/rw/RWB.NSF/db900SID/LZEG-6X4MWH?OpenDocument> (accessed 12 May 2009).
18. "The Costs to the Taxpayer Are Enormous". This report identifies nineteen Katrina contracts collectively worth US$8.75 billion that have been plagued by waste, fraud, abuse or mismanagement. See United States House of Representatives Committee on Government Reform — Minority Staff Special Investigations Division, *Waste, Fraud, and Abuse in Hurricane Katrina Contracts*, August 2006 <http://oversight.house.gov/ Documents/20060824110705-30132.pdf> (accessed 12 May 2009).

Similarly, according to "a January 2005 audit report from Stuart W. Bowen, the government's special inspector general for Iraq reconstruction, ... [U.S. authorities] failed to account for $8.8 billion given to Iraqi ministries". Allegedly, 360 tonnes of cash "were taken from the Federal Reserve in New York, loaded onto wooden pallets and put on cargo planes that were flown into Baghdad. Representative Waxman claimed [that] $12 billion dollars were sent to Iraq between May 2003 and June 2004 and is unaccounted for by the

U.S. government"; see Jennifer Parker, "Waste in War: Where Did All the Iraq Reconstruction Money Go?" 6 February 2007, ABC News <http://abcnews.go.com/ Politics/story?id=2852426> (accessed 12 May 2009).

19. The following text draws heavily, though not uniquely, on the TEC Synthesis report: J. Telford and J. Cosgrave, *Joint Evaluation of the International Response to the Indian Ocean Tsunami: Synthesis Report* (London: Tsunami Evaluation Commission, 2006).

3

LINKING RELIEF, REHABILITATION AND DEVELOPMENT (LRRD) TO SOCIAL PROTECTION
Lessons from the Early Tsunami Response in Aceh[1]

Ian Christoplos and Treena Wu

A GOOD START, BUT WHAT ABOUT DEVELOPMENT?

The tsunami had an immense impact on development processes, conflicts, patterns of risk and poverty in affected areas, as did the subsequent relief and reconstruction efforts. This chapter looks at how affected populations in Aceh have coped with the disaster from a social protection perspective. It considers how they have continued to manage their own welfare both with and without aid. Specifically, it focuses upon issues of poverty alleviation, equity, and the linking of relief, reconstruction and development (LRRD) in the aftermath of the 2004 tsunami. As part of this exploration, questions are asked about whether the reconstruction and development efforts have buttressed or compromised the social protection mechanisms that people already had in place at the family and community levels prior to the disaster.

Conceptually, "social protection" comes from the European welfare state design that recognizes three pillars of welfare: private, government and market. Resources move among these three pillars in accordance with the welfare needs of individuals and families. Such a design recognizes and responds to changing levels of societal development and consequences that create risks. In the case of Aceh and the rest of Indonesia, social protection needs to factor in the many unpredictable man-made and natural disasters faced, as part of efforts to better link these pillars.

There is a growing realization that, from the perspective of the economically disadvantaged, shocks are not abnormal. Disasters and hazards, large and small, are an ever-present threat that profoundly affects livelihood strategies. In recognition of this, social protection has become a central issue in the development debate in many countries, even though the term means different things to different people.[2] Social protection may encompass social assistance schemes, social security, social funds, cash transfers and other structures that provide a safety net for those at risk of becoming destitute, or a safety net to provide subsidized support for those struggling to rebuild their lives in the aftermath of trauma. These may be either formal (state-managed) or informal, such as the community self-help mechanisms that were mobilized directly after the tsunami in many locations. Formal mechanisms are related to government social security schemes for citizens: old-age pension, old-age savings, national health insurance, work-injury insurance and death benefits for survivors of deceased workers. Informal mechanisms are concerned with how individuals, households and villages manage their own welfare and protect themselves against social and economic risks. One of the most important mechanisms for this in Indonesia is short-term circular migration, as discussed in more detail in this volume by Mahdi.[3] Neither formal nor informal social protection mechanisms are generally an immediate concern of aid agencies, particularly humanitarian agencies, but if recovery efforts are ultimately to reduce vulnerability, they must become part of the equation.

A disaster, as defined by the UK Department for International Development (DFID), is an event that temporarily overwhelms the above-discussed formal and informal structures, thus requiring a humanitarian response that steps in to provide a basic package of social security. A key conceptual challenge in LRRD is to decide how temporary such interventions should be, and how to smoothly manage the shift back to reliance on local social protection institutions. However, in many post-disaster situations, most aid actors lack detailed historical and context-specific knowledge that is critical for understanding what kinds of interventions might eventually prove sustainable. In the case of Aceh, ignorance of livelihoods, community structures and *gampong* social networks impinged heavily on social protection. The discord between aid providers and society has been magnified through project-driven processes that have failed to recognize how affected persons and communities managed their own welfare in the past. This was compounded by ignorance concerning the capacity of local government authorities to manage these interventions in the development phase.

To illustrate, during the first six months after the tsunami, many agencies found available land and built temporary shelters or started building permanent housing. Because of the absence of dialogue between the aid agencies and internally displaced persons, the construction carried out did not factor in mobility patterns for livelihood activities, or the full implications of customary and formal land laws (see Fitzpatrick in this volume). Before the tsunami, people had to protect their families and livelihoods alone without external intervention. A range of informal social protection mechanisms, such as localized patterns of migration, were used. Through their information networks, individuals would travel long distances daily from their homes to the closest town to take on some work.[4] Furthermore, community-based volunteer efforts, such as *gotong-royong*, were fundamental in assessing the needs of individual households and carrying out livelihood activities collectively to meet these needs. Many such mechanisms were circumvented by aid and reconstruction efforts.

The importance of government and community ownership of the recovery process is formally acknowledged by almost the entire aid sector, but during the first year of the tsunami response, there were frustrations and delays in anchoring tsunami response in Acehnese institutions. Genuine LRRD requires aligning programming with the policies, capacities and actions of national actors, be they governmental, civil society or the affected populations themselves. The weaknesses in national and local institutions were and still are considerable, so alignment can be expected to be a protracted process. There are indications that this alignment is now improving, but in some areas significant damage has already been done due to poaching of staff and insufficient attention to pre-existing policy frameworks and evolving social and political dynamics.

THE CORE CHALLENGE IN LRRD: ALLEVIATING BOTH TRANSIENT AND CHRONIC POVERTY

Disasters can generate both lasting and temporary forms of poverty. Post-disaster aid modalities and changes in the political and economic context in Aceh have affected both chronic and transient poverty. Effective LRRD manifests itself in a judicious balance of efforts to address both. It is a matter of bringing together the different principles of humanitarian and development response and ensuring that both are respected and acted upon.

Given the extent of resources available in the tsunami response, it would be unwise to conceptualize this as an either-or question, of whether to focus on humanitarian needs or to invest in employment opportunities for those made chronically poor by the disaster. There is enough money to do both. The LRRD question is whether efforts have been designed that can combine humanitarian and developmental interventions in a balanced and appropriate way that are aware of and sensitive to local concerns.

Before the tsunami, chronic poverty was widespread in Aceh. According to the 2003 estimates of Badan Pusat Statistik, almost thirty per cent of the population in Aceh was living below the poverty line, against the national average of 17.4 per cent. At the time, Aceh was the third-poorest region in Indonesia, despite being endowed with natural resources that are mainly extracted for use at the national level. The tsunami resulted in extreme levels of transient poverty, as a large group of those who were moderately well off were left destitute. Despite some claims to the contrary, the tsunami not only affected the poorest of the poor, but almost certainly pushed more people into this category. It is important to consider how much of the destitution created by the tsunami can realistically be reversed through short- to medium-term rehabilitation projects based on asset replacement, and how many people can be said to have entered the ranks of the chronically poor.

Evidence from other disasters has shown that poor peoples' coping strategies in the face of disasters can lead them into poverty traps, where their depletion of assets leads to long-term destitution. Income shocks, destruction of homes and productive assets, reduced consumption/nutrition and stress sales of productive assets are interlinked.[5] There is no consensus regarding how to best avoid this weak resilience through relief and rehabilitation. More developmental interventions may be appropriate, but it is also unclear how such modalities should be designed. Weak resilience is a concern for LRRD efforts, but is primarily a challenge that relates to the poverty alleviation efforts of the affected countries.

A significant proportion of the transient poverty caused by the tsunami has now been alleviated either due to aid or to the efforts of the affected populations themselves. Some has not. The loss of productive agricultural land, facilities with which to run commercial enterprises, and jobs that are created through commercial enterprise cannot all be remedied with short-term livelihood projects. Those who rely on these enterprises have not generally been reached by the tools of relief and rehabilitation.

Social protection is intended to be the cornerstone of efforts to increase the resilience of the poor who face recurrent shocks to their livelihoods and well-being. A fundamental question is determining where disaster response ends and where the responsibilities of the state, civil society and local communities for basic social protection begins. This is especially the case for those who have less chance of rising from their tsunami-related destitution and are now chronically poor. Chronic poverty has increased after the tsunami, due to loss of or loss of access to:

- Productive agricultural land
- Human resources (e.g., family labour, skills)
- Land and housing rights, *adat* or formal title
- Social capital that comes from the fracturing of *gampong* units and kinship ties
- Fixed capital (especially for women with home-based enterprises)
- Financial capital (combined with increased debt)
- Pensions and other social protection measures due to loss of identity papers
- Public service institutions

RIGHTS AND RESPONSIBILITIES FOR SUSTAINABLE POVERTY ALLEVIATION AND SOCIAL PROTECTION

"Rights-based approaches" are conceptual frameworks for aid programming that relate to the fundamental questions of how to combine humanitarian and developmental frameworks for addressing poverty. Humanitarians and development actors tend to have very different perspectives on what constitutes a "right" in different contexts and of the relative importance that should be given to sustainability versus speed in upholding "rights". Differing perspectives on sustainability are not the only source of confusion across the LRRD divide. "Standards" have become the most common vehicle for efforts to uphold rights. However, different agencies and phases of response raise the following questions about whose "standards" are relevant:

- Should one follow the international standards of the humanitarian community (e.g., Sphere)?
- Should national guidelines for poverty alleviation, e.g., Poverty Reduction Strategy Papers (PRSPs) and attainment of the Millennium Development Goals (MDGs) steer planning?
- Should local people be supported to define their own standards through participatory approaches?

The importance of adherence to one or more of these standards is frequently invoked in tsunami response plans, but rarely is there discussion of whether or not programming can contribute to a sequencing of efforts in response to the relative prominence of different "rights" in different contexts. Adherence to standards has sometimes been promoted as an end in itself, as opposed to being a means for obtaining a range of desirable outcomes. Provision of water has taken precedence over solid analysis of how to ensure sustainable access to water. Being responsive to standards alone can be an excuse not to take contextual needs into consideration. When barracks were being built as temporary shelters, many projects were solely focused on the shelter construction standards set in Sphere. Very few assessed whether these barracks had a negative connotation for families who had suffered through the conflict, or were culturally suitable for Acehnese use.

A starting point for defining what a rights-based approach means in the tsunami LRRD response may be to see if efforts are becoming part of a framework of social protection, whereby protection from shocks and reduction of risks and vulnerabilities are seen to be a responsibility for the state, international providers and local communities. In other words, is the protection of vulnerable people considered an integral part of commitments to alleviate poverty, or are shocks perceived as an aberration to be addressed separately from ongoing poverty strategies by discrete humanitarian projects?

The humanitarian-oriented aid discourse in Aceh tends to refer to high levels of destitution as indicating a need to continue relief programmes (or even as indicating a continuing right to relief). This mindset has proven an obstacle to LRRD. There is a large population of chronically poor who are not likely to be lifted out of poverty by yet another short-term project. They will only be effectively supported if sustainable social protection structures are (re)established. Despite the efforts of some agencies, such developmental approaches to shocks and destitution have not yet been scaled up to meet the new challenges.

In scaling up, one of the main challenges is how to link efforts with different institutions at local and national levels. Across Indonesia, local government capacity is uneven, and is now a major binding constraint to poverty reduction. About one-third of total public expenditure is allocated and spent at the district level.[6] While this is a sign that decentralization is underway, the problem is that many local governments are facing difficulties in planning, budgeting and executing this spending.[7] To Aceh's credit, public works interventions have succeeded in simultaneously addressing the need to provide an income to households as well as rebuilding urgently needed infrastructure.

Social protection through public works investments and decentralization is particularly appropriate since Aceh has economic resources to ensure the survival of its own population. There is no reason to assume or accept that the chronically poor should be seen to be wards of the international community for an extended period of time after the tsunami. Furthermore, the strong response from community members, local businesses and civil society in the immediate aftermath of the tsunami demonstrated the strength of informal social protection mechanisms. These responses provide further evidence of how the Acehnese look after each other, and is consistent with the sociological phenomenon of Indonesian life in general that centres on protecting the village.[8] If humanitarian agencies stay on too long, this could result in damage to these mechanisms as people grow accustomed to seeing disaster victims as being the responsibility of the international community. Such a development would indicate that great harm has been done to informal social protection structures and indigenous forms of disaster preparedness and response.

The implications of these issues are particularly notable with regard to defining an appropriate role for the humanitarian sector in addressing the special needs of the elderly, the disabled and other groups that are not effectively reached by standard programming modalities. The TEC LRRD reviews found that very few agencies have paid attention to the rights of these affected populations for assistance that meets their needs and builds on their capacities. Emergency assistance should address the unique, acute needs resulting from their loss of shelter and other assets. This includes adapting shelter and resettlement plans to reflect their needs, targeted health services and ensuring that the shift to cash-for-work modalities does not exclude support to those who cannot participate in these programmes. An equally important question is whether aid helps or hinders the rebuilding of the social protection structures on which they will rely in the future.

OPPORTUNITIES AND OBSTACLES TO LEARNING ABOUT POVERTY AND VULNERABILITY

Responding to poverty demands an understanding of its multidimensional nature. It also requires an understanding of how affected countries and different groups of poor people have historically dealt with poverty. This knowledge is limited among many of the agencies that are involved in the tsunami response. The TEC evaluations repeatedly note that the aid community has not drawn on the considerable experience existing in the affected countries in dealing with different forms of crisis transitions in the past.[9]

In the case of Aceh, which suffered from at least thirty years of conflict prior to the tsunami, this ignorance is exacerbated by the paucity of information that is readily available about Acehnese culture and society. Even the Indonesian National Socio-economic Survey's (SUSENAS) information about Aceh is very limited relative to other parts of Indonesia. For example, many of the aid agencies arrived in Aceh without understanding its deep Islamic roots and entrepreneurial tradition; this has strong implications for how the delivery of aid was designed and implemented to respond to poverty.

The people of Aceh were not passive in the face of inappropriate aid "standards". In numerous surveys and participatory planning exercises, the response from Acehnese society to the so-called aid tsunami has been to stress the importance of their cultural inheritance as a starting point for reform rather than an obstacle to reform. In a community survey with people from *kabupaten* Pidie who had resettled in Malaysia,[10] their leaders cited that a priority when fleeing to Malaysia was to ensure that all historical documentation about Aceh was protected. Such documentation helps to trace the different dimensions of poverty. By identifying and understanding the dimensions of poverty and their interactions over time in the context of Aceh, the issues of poverty reduction can be better addressed.

Using this same *kabupaten* to illustrate what the aid community missed in its poverty analysis, this survey found that migrant workers known as *Tenaga Kerja Indonesia* (TKI) and *Tenaga Kerja Wanita* (TKW) working in Malaysia actively support poor households in Pidie. Many aged people in particular receive some form of income support from migrant workers. But this mostly went undetected by the aid community when assessing the needs of this vulnerable group.

Of course, emergencies are not ideal opportunities for learning about the dynamics of poverty. Operational actors tend to give higher priority to "action" than to critical analysis of the relevance of their work. This deficiency should not, however, be accepted as inevitable. One in-depth study of World Bank operations in Indonesia noted that, up until the late 1990s, poverty analysis was weak, but "the [economic] crisis of 1997 made it evident that this was not enough.... This manifested itself in a dramatic shift in the country portfolio as the Bank scrambled to find instruments to deal with poverty and crisis".[11] Because of this crisis, Indonesia became one of the countries where World Bank operations underwent a fundamental refocus as efforts shifted to decentralization and social safety nets. But even within these efforts, Aceh continues to be distinctly different from other parts of Indonesia. It has its own dimensions of poverty that require long-term, local-level research.

Attempts to relate current tsunami programming to past experience and current development policies were few and far between. In the first year of the response, there were almost no references to the Millennium Development Goals (MDG) in international tsunami-related programme documents and reporting (thirteen citations out of over 24,000 documents collected by the TEC). Mentions of Poverty Reduction Strategy Papers (PRSP) were even rarer (one citation found). These poverty alleviation strategies and objectives may have significant impact on whether short-term projects can contribute to longer-term efforts, but it is doubtful that the majority of agency staff were aware of national plans for poverty alleviation.

The issue is not just one of relating LRRD responses to pre-existing poverty alleviation strategies, but also of ensuring that these poverty alleviation strategies are reconsidered in light of the LRRD processes of the aid community and the affected populations. Organizations may need to reconsider their roles in addressing chronic poverty, based on what has been learnt about the vulnerabilities revealed by the tsunami. There is little indication that learning about poverty has been a significant feature of the post-tsunami period.

EQUITY, A MYRIAD OF DEFINITIONS

Equity and poverty alleviation are two different things, but they are often conflated in discussions of LRRD. Some of the poverty resulting from the tsunami can be alleviated without addressing structural inequalities. Some cannot. Furthermore, aid can be provided in an equitable manner without contributing to the alleviation of poverty. Indeed, most humanitarian assistance does not aim to alleviate poverty but strives towards equity in order to ensure the basic survival and dignity of all. The chronically poor may be especially targeted, but they are not the exclusive target group.

A significant feature of LRRD in the early tsunami response was the assumption that, if assets were distributed in an equitable manner, poverty would thereby be alleviated. For example, provision of small boats in the name of equity was seen as a sufficient guarantee that poverty would be alleviated. Analysis of how the fishing industry could sustainably contribute to poverty alleviation was not part of programme design. More detailed analyses of the political economy and the biophysical resource base of fisheries would have been necessary if the two objectives of equity and poverty alleviation were to be effectively combined, but these have been rare.

It is important to observe that the term "equity" has been applied in a myriad ways, implicitly and explicitly, in the tsunami response. It can mean giving priority to any one or more of the following categories and goals:

- Those most affected by the disaster
- Impartially allocating assistance to those affected by the tsunami and those affected by the conflict according to relative need
- Looking beyond the impacts of the tsunami per se, to support those most in need, including conflict-affected populations and the chronically poor
- Those least capable of recovering without external support
- People who are disadvantaged due to age or physical capacities
- The poorest
- Ensuring equal access to aid for men and women
- Those enterprises that are most likely to generate sustainable employment
- Equitable distribution of aid resources *within* recipient communities
- Equitable distribution of aid resources *among* affected communities
- Equitable distribution of aid resources *among gampong* units

There are overlaps between these priorities, but none of them are congruent. The choices between these priorities above are not just issues of policy. They are also a reflection of political struggles, cultural norms and perceptions of how a disaster response should be situated in broader development processes. The choices among these priorities should be part of a larger dialogue in which the Acehnese define their own concepts of equity and how they are related to social protection.

Although equity and its operational function of "targeting" concern many organizations, there are few agency policies that disaggregate these objectives and notions of equity in the tsunami context. Programming decisions reflect a combination of the above definitions without clearly discerning which type of equity is more or less important. One notable exception is a UN High Commission for Refugees (UNHCR) discussion paper that proposes clear principles for addressing the needs of different types of the internally displaced and other disadvantaged groups.[12]

Most agencies assume that the poor are those who were most affected by disasters and that a focus on asset distribution to the poorest will thereby contribute to both poverty alleviation and rehabilitation for those most affected. This is further underpinned by the notion that direct support to own-account production is inherently more equitable than indirect support through reinvestment in larger enterprises. In light of the findings behind many current bi- and multi-lateral poverty alleviation policies,[13] these deeply flawed assumptions need to be questioned. This belief that supporting own-account production in order to promote equity as an inherently appropriate basis for poverty reduction has been referred to as the "yeoman farmer fallacy".[14] It is an assumption that has been questioned for many years in development circles, but is largely undisputed among the international NGOs involved in the tsunami response. The reason for this is that the humanitarian assistance modalities of the NGOs focus almost exclusively on *directly* assisting those in greatest *need*. The creation of opportunities for employment does not enter the equation of those agencies "doing livelihoods". Some agencies have explicitly tied themselves to yeoman farmers and fisherfolk by stating that they respond only to need, and not to losses.[15]

This is a fully justifiable response in a purely humanitarian action. But in rehabilitation and development issues of sustainability, public finance of basic social services, alignment with national policies and congruence with local norms must be addressed alongside the "humanitarian imperative". Poverty alleviation inevitably becomes an issue with political and ideological dimensions. A comprehensive alignment between poverty alleviation policies and LRRD in Indonesia would seem to require a dual strategy that responds to both needs (i.e., through social protection) and also supports the expansion of opportunities for employment. Such a broader developmental perspective would require complementing the needs-driven approach with measures to support the replacement of *losses* by medium and large enterprises, which may in turn *indirectly* (but more effectively and sustainably) reach those who have lost their productive resources by creating employment opportunities. It is these losses that reduce economic activity which may have the greatest impact on the poor, especially in a highly entrepreneurial society such as Aceh.[16] Such measures may also re-establish a tax base with which to finance government services. It should be noted that the three countries/regions that went through the most massive LRRD operations in the last ten years (Bosnia Herzegovina, Kosovo and East Timor) are all currently confronted with catastrophic unemployment and public finance crises. These examples suggest that the aid industry has missed the target in addressing the need to create sustainable employment opportunities in the past. There is no indication that lessons regarding LRRD impacts on poverty alleviation are actively being sought from these countries to apply in the tsunami response.

Gender equity issues have not been ignored, but the prevailing methods of livelihoods support have limited agencies' ability and readiness to turn gender awareness into more equitable programme modalities. "Gender focal points" abound in NGOs, but their broader influence on programming can be seriously questioned. Targeting related to gender equity demands more than just ensuring that women are not missed in beneficiary lists. It requires an understanding of what is happening within households. From the 1960s to 1990s, Indonesian society as a whole has stressed the importance of *ibu*-hood and the role of women in home production.[17] Conceptually, *ibu*-hood refers to women as being good wives and mothers, meaning that their role in the public domain is subject to many demands. Since the tsunami, community surveys carried out in male-dominated groups indicate that many households give priority to restoring the male breadwinner's livelihood. Allocation of resources automatically goes in the direction of the men with little investigation of the women's needs and, more importantly, how their roles in the public domain were affected. How does a woman struggle between looking after her children and looking for clean water in an IDP camp? Why does the water tanker managed by a water and sanitation project drive by the camp only after lunch? How is the mother going to find clean water to cook lunch for her family? Does a woman have economic livelihood activity?

LESSONS FOR THE FUTURE

LRRD is not a set-piece process. It demands knowledge of the political economies of the countries and the social complexities of communities affected by the disaster. It also demands capacity and readiness to learn at field level. Agencies have not been proactive in building their contextual knowledge and relationships with local institutions and civil society in Aceh. The act of building knowledge and relationships has too often been limited to creating mechanisms for disbursing funds at a faster rate for projects. This does not bode well for the emergence of constructive relationships where the links to development and sustainable social protection are found through dialogue with local institutions. The unprecedented quantity of funding available has carried with it a tendency to worry more about how an activity will appear "back home" than about its relevance for affected populations.

The overall implication for the future is that there is a need to break out of the "project-focused" paradigm of aid provision. This is in order to acknowledge that the most significant links between relief, rehabilitation and the more long-term establishment of social protection institutions are those that are made by affected populations themselves and by the national public and private institutions on which they depend for jobs, services and human security. The people affected by the tsunami are getting on with their lives regardless of the sometimes chaotic and ill-conceived actions of the aid community. Improving LRRD programming is thus not a matter of agencies becoming better at "doing livelihoods" or even building houses. It lies instead in deeper analysis of how "our" meagre efforts can better contribute to supporting "their" LRRD projects.

Attention to "their" LRRD projects leads inevitably to greater engagement in micro- and macro-political processes. This creates a well-justified unease among some humanitarian agencies that are concerned about how to maintain adherence to the humanitarian principles of neutrality, impartiality and independence. Indeed, effective LRRD demands close engagement with local institutions, with a consequent loss of independence. Weakened adherence to some aspects of humanitarian principles can nevertheless be balanced by political savvy, clarity of commitment and contextual awareness, so as to ensure impartiality and neutrality in conflict situations and amid political efforts to influence resource flows. The predominance of staff with no experience in Aceh raises concerns that they may not have the necessary skills to manoeuvre amid the micro-political realities of LRRD.

These concerns point towards two overall conclusions. First, for LRRD to become more effective, the aid industry needs to greatly increase its capacities for engaging with local and national development processes. This is reliant on a humble acknowledgement of the enormity of the tasks of reconstruction and a more proactive search for ways to work constructively with institutions at national and local levels. Second, many agencies evidently lack the capacity to take on sizeable LRRD engagements in an effective manner. National authorities and donors should work together to ensure that agencies are not allocated responsibilities that are reliant on skills that they obviously cannot muster. Our research in Aceh and elsewhere leads us to suggest the following points for making LRRD programmes more sustainable:

1) LRRD must be more firmly rooted in the local context, driven by locality, *adat* and past histories, and related to national-level processes. A bridging of the current divide between aid programming and the initiatives of affected populations will require a reconsideration of how aid contributes to or hinders the decentralization agendas of national authorities, local officials, NGOs, businesses and the affected populations.

2) Links between relief and rehabilitation have been achieved, but greater attention needs to be paid to the implications of programming for longer-term development. Foreign aid has not been the only, or even the primary, motor for restarting economic activities, so neither full credit nor blame should be attributed to the aid industry. Nonetheless, the viability of

many of the livelihoods supported by aid programming is questionable. Shelter, housing and land rights have frequently been addressed in a narrow perspective, without sufficient concern for the functionality of the communities being rebuilt and created. Also, the capacity of local government institutions to manage initiatives over time has to be taken into account.

3) Poverty alleviation interventions need to be better related to ongoing poverty alleviation trajectories. Effective LRRD manifests itself in a judicious balance of efforts to tackle both chronic and transient poverty. Progress has been rapid in alleviating much of the transient poverty that was created by the tsunami. However, there are a significant proportion of people whose tsunami-related destitution has now effectively placed them in the ranks of the chronically poor. They are unlikely to be helped by small asset replacement initiatives. Their needs are best addressed by economic development underpinned by social protection. The public provision of social safety nets needs to run in parallel with existing informal welfare mechanisms in the family and community.

4) More consideration needs to be given to reducing risks of natural disasters, conflict and other factors, and anchoring such strategies within national social protection structures. Despite additional international attention and funding for early warning, risk reduction has not been fully mainstreamed in recovery programming. There is a need for deeper and more evidence-based assessment of the impacts of aid programmes on environments and natural resources. Given the prevailing risks, there is a need to consider how national structures can re-shoulder responsibilities for social protection (e.g., the use of national-level-funded disaster insurance) to deal with various forms of shocks from natural hazards, conflicts and other factors. This is related to how resources are allocated between the three pillars of welfare. Aid needs to be refocused so as to support governments as they reassume responsibility for ensuring the safety, survival and dignity of their citizens.

5) Links to the LRRD efforts of affected populations should be improved through strengthened information flow. Disaster-affected people need information about the aid they will receive so they can decide how best to rebuild their lives and livelihoods. It is more important than "participation", since participation in aid projects is secondary to the efforts of affected populations to get on with their own LRRD projects. The affected populations in Aceh have not received sufficient information and they are justifiably angry, frustrated and confused. Provision of better information can make a modest but important contribution to

strengthening the clout of affected populations in influencing the LRRD agenda.

6) LRRD can be best served by greater transparency and institutional accountability about who is able to do what, and when they are able to do it. The problems that have emerged in LRRD often relate more to agencies having promised too much than to them having done too little. Agencies, donors and government authorities have felt pressured to make commitments that are far beyond what they can actually accomplish. Criticism should therefore not necessarily be directed at their failures to achieve these objectives, but rather at the ways that these claims have led to unfulfilled promises to affected populations and dysfunctional shortcuts in development planning.

SUMMARY OF CONCLUSIONS

To conclude this chapter, the lessons learned from Aceh have important implications for other post-disaster situations. LRRD response to any disaster situation has to be well thought out despite the urgency at hand. The response has to be made with the expectation that there will be links to the development and social protection processes that were already occurring in the disaster-hit area. This requires that humanitarian actors enter into dialogue with government at the local and national levels.

At the local level, the payoff for the international community is that eventually the affected population can assume full ownership of the reconstruction process. Sustainability must reflect what people at the local level would like to sustain, which in turn implies that efforts should reflect the political dynamics of a given population. At the national level, there must be alignment between the aid resources allocated to the disaster-affected area and national-level management of resource use. This has implications for the scaling up of aid projects to the level that is financed by public funding. This is particularly important with respect to institutionalized social protection such as old age care and transfers to the poor such as *Bantuan Langsung Tunai*. However, as shown by the experience in Aceh, aid agencies responsible for relief and rehabilitation seldom have the capacity to enter into dialogue and to remain engaged in the medium term. They are unable to bring relief aid assistance into the national planning agenda, particularly poverty alleviation trajectories. As a consequence, relief and rehabilitation work in post-disaster situations tend to resemble more what was done in the last major disaster rather than reflecting the unique needs of the affected area.

Currently, the Government of Aceh is taking ownership of its political and governance processes. As such, it is likely that foreign aid now being disbursed will contribute more effectively to social protection for the vulnerable than earlier programming. This shows that links from relief and rehabilitation to the development agenda do eventually emerge. The question is whether they emerge before the recovery funds have already all been spent.

Notes

1. This chapter expands on a synthesis review of "Links between Relief, Rehabilitation and Development in the Tsunami Response", undertaken as part of the work of the Tsunami Evaluation Coalition (TEC); it is written in the context of the social, political and governance spheres in Aceh. The TEC study on linking relief, rehabilitation and development was financed by the Swedish International Development Cooperation Agency (SIDA). The views presented here are those of the authors and do not necessarily represent the views of SIDA or the TEC.

2. A. Barrientos, D. Hulme and A. Shepherd, "Can Social Protection Tackle Chronic Poverty?". *European Journal of Development Research* 17, no. 1 (2005): 8–23.

3. G. Hugo, "Circular Migration in Indonesia", *Population and Development Review* 8, no. 1 (1982): 59–83.

4. T. Wu, "The Role of Remittances in Crisis: An Aceh Research Study", in *Remittances in Crises*, edited by Paul Harvey and Kevin Savage (London: Overseas Development Institute, Humanitarian Policy Group, 2007).

5. M. Carter, P. Little, T. Mogues and W. Negatu, "Shocks, Sensitivity and Resilience: Tracking the Economic Impacts of Environmental Disaster on Assets in Ethiopia and Honduras", Economic Working Paper Archive EconWPA, 2005 <http://econwpa.wustl.edu:80/eps/dev/papers/0511/0511029.pdf> (accessed 12 May 2009).

6. World Bank, *Making the New Indonesia Work for the Poor* (Jakarta: World Bank, 2006).

7. Ibid.

8. Y. Yuliati and M. Purmono, *Sosiologi Pedesaan* (Yogyakarta: Lappera Pustaka Utama, 2003).

9. J. Cosgrave, *Synthesis Report: Expanded Summary — Joint Evaluation of the International Response to the Indian Ocean Tsunami* (London: Tsunami Evaluation Coalition, 2007).

10. Wu, "The Role of Remittances in Crisis: An Aceh Research Study".

11. A. Bebbington and A. Barrientos, "Knowledge Generation for Poverty Reduction within Donor Organizations", GPRG Working Paper Series 023, Manchester: Global Poverty Research Group (GPRG), 2005 <http://www.gprg.org/pubs/workingpapers/pdfs/gprg-wps-023.pdf> (accessed 12 May 2009).

12. UN High Commissioner for Refugees (UNHCR), "The Internally Displaced in Sri Lanka: Discussion Paper on Equity", IDP Working Group in Sri Lanka, 2005 <http://www.reliefweb. int/library/documents/2005/unhcr-lka-1dec.pdf> (accessed 12 May 2009).

13. SIDA, "Improving Income among Rural Poor: Strategic Guidelines for Sida Support to Market-Based Rural Poverty Reduction", SIDA, Department for Natural Resources and the Environment, Stockholm, 2004 <http://www.sida. se/shared/jsp/download.jsp? f=SIDA4088en_Rural+poor_web.pdf&a=3260> (accessed 12 May 2009).

14. J. Farrington and A. Bebbington, "From Research to Innovation: Getting the most from Interaction with NGO's in Farming Systems Research-Extension", invited paper for the International Farming Systems Research-Extension Symposium, Michigan State University, 14–18 September 1992.

15. Oxfam, "Targeting Poor People: Rebuilding Lives after the Tsunami", Oxfam Briefing Note, 2005 <http://www.oxfam.org.uk/what_we_do/issues/conflict_ disasters/downloads/ bn_tsunami_women.pdf> (accessed 12 May 2009).

16. E. Clay and C. Benson, "Aftershocks: Natural Disaster Risk and Economic Development Policy", ODI Briefing Paper, Overseas Development Institute, London, 2005 <http://www.odi.org.uk/publications/briefing/bp_disasters_nov05. pdf> (accessed 12 May 2009).

17. A. Niehof and F. Lubis (eds.), *Two Is Enough: Family Planning in Indonesia under the New Order (1968–1998)* (Leiden: KITLV, 2003).

4

CULTURAL HERITAGE AND COMMUNITY RECOVERY IN POST-TSUNAMI ACEH

Patrick Daly and Yenny Rahmayati

Our experiences in Aceh lead us to believe that a surprising amount of the reconstruction and development agenda has failed to address the cultural and historical dimensions of social recovery. In spite of all the meetings, coordinating sessions and public statements about interagency cooperation, it is impossible to find a commonly-accepted definition of what 'recovery' entails or should look like. It is difficult to imagine such resources could be allocated and spent[1] without a clearly-defined end-game, but unfortunately this is an endemic problem in many post-disaster situations[2] (Bennett et al. 2006; Telford and Cosgrave 2006). The evidence from Aceh suggests that this is specially a concern when there are large numbers of external organizations involved in aid and reconstruction processes.

Naomi Klein's influential book *The Shock Doctrine* draws attention to the cynical and opportunistic behaviour often accompanying post-conflict and post-disaster reconstruction processes. She builds a powerful argument that government and corporate interests exploit the aftermath of large-scale social trauma for political and/or economic gain, and in some cases initiate or encourage trauma (Klein 2007). While there are certainly cases where relief and development aid is manipulated to achieve political, economic or social goals, we want to argue in this chapter that practices which target social transformation that fall outside pre-existing social and cultural contexts can impede the recovery of traumatized communities, even when it is brought about through the interventions of well-meaning individuals and organizations. Our research focuses upon cultural and social mechanisms for community recovery, and how these are related with the material world. We argue that there are aspects of post-disaster recovery that are contingent upon reconnecting with familiar cultural and social practices, which in turn are intimately connected with the built environment. Aid and reconstruction efforts that further remove people from familiar physical and social contexts run the risk of pulling them away from the basic community infrastructure that is necessary for recovery. We demonstrate that in Aceh this essential reconnection has been circumvented by national and international organizations that have been largely ignorant or dismissive of local cultural and social practices. This is supported by evidence from field surveys conduced in Aceh, and reinforced by literature from a range of social sciences.

FROM 'BUILDING BACK BETTER' TO RECONNECTING WITH THE CULTURAL PAST

We will rebuild Aceh and Nias, and we will build it back better...
— Susilo Bambang Yudhoyono, President of Indonesia (2005)

A stated philosophy and widely-branded slogan promoted by the Agency for the Rehabilitation and Reconstruction for Aceh and Nias (*Badan Rekonstrusi dan Rehabilitasi di Aceh dan Nias* BRR), the Indonesian organization charged with overseeing and coordinating the aid efforts in areas damaged by the Tsunami, is 'Build Back Better'. This phrase is prevalent in government and NGO literature dealing with post-Tsunami reconstruction, and, with few exceptions, has not been challenged.[3] At first glance it is difficult to disagree with building back *better*, as it is ostensibly well-intentioned. However, we feel that in the drive to build back better, some factors that are vital for achieving longer-term community recovery are overlooked. In this chapter we don't want to get bogged down evaluating whether or not agencies were able to 'build back better' in a literal case-by-case basis, but rather to critically challenge the usefulness of the concept in post-Tsunami Aceh.[4] This allows us to more fruitfully discuss the complicated relationship between change and recovery in post-disaster environments, and the importance of cultural practices and historical narratives within this.

Admittedly by most criteria, following decades of conflict, isolation, and troubled economy, Aceh had significant problems at the time of the Tsunami (Reid 2006). Aceh long resisted different waves of European colonization, often involving extensive fighting and periods of occupation by the Portuguese followed by the Dutch. This strife continued after Indonesian independence when factions in Aceh, most notably *Gerakan Aceh Merdeka* (GAM), waged a low intensity separatist campaign against Jakarta. Aceh was under a state of marshall law at the time of the tsunami, and hostilities were not formally ended until the signing of the Helsinki Agreement in August 2005 which effectively ended Acehnese aspirations for full independence, and brought a period of peace and stability.

However, Aceh's problems do not exclude it from being endowed with a rich array of social and cultural practices, and a long and proud historical consciousness,[5] all of which are key elements within processes of community recovery. We see *build back better* not just as a negative statement about Aceh and its cultural and social institutions prior to the Tsunami, but also as part of a globally-accepted justification for imposing an externally-driven top-down reconstruction agenda.[6] The call for change explicit within this directive does not have its origins within Aceh, and has been widely translated on the ground to include both the physical reconstruction of buildings and communities, and also programmes that focus on social transformation to create 'better' living conditions and social opportunities in Aceh.[7] This comes despite all the talk within the NGO community about their roles in the process, and the need to pursue 'participatory' and locally-sensitive reconstruction. It has become almost an assumption within the reconstruction industry that 'windows of opportunity' afforded by disasters should be seized to usher in a wide range of economic and social development, clearly seen in the popularization of LRRD — Linking Relief, Reconstruction and Development (Christoplos 2006).

Our experiences in Aceh and other traumatized situations, coupled with extensive study of the literature from a number of disciplines dealing with post-disaster recovery lead us to counter 'build back better' with 'reconnecting with the cultural past' as another lens for conceptualizing post-trauma relief and reconstruction projects. This is based upon our understanding of community recovery as re-establishing as best as possible the social trajectory and momentum that existed within a community prior to a disaster, to the point where communities can manage the longer-term affects of devastation and trauma within frameworks of stability and change defined internally. We believe that there is a greater possibility of sustainability if programmes do not exceed the expectations, capacities, and cultural sensibilies of those who have to manage and live with the consequences of such efforts, long after external support systems have left.

It is widely accepted in relevant social science and psychology literature that there are latent capacities within individuals and communities that allow them to deal with stress and trauma (Brickman et al. 1982; Omer and Alon

1994; Rich et al. 1995; Norris and Kaniasty 1996; Oliver-Smith 1996; Gilbert and Silvera 1996; Gist and Lubin 1999; Bonanno 2004; etc.). A review of this literature suggests that people and communities typically are endowed with powers of resilience that enables them to respond to and recover from trauma (Bonanno 2004; Bonanno and Keltner 1997; Cardena et al., 1994) and that outside assistance has to be very mindful of interrupting or co-opting indigenous response mechanisms (Gilbert and Silvera, 1996; Oliver-Smith 1996).[8] Furthermore, it has been amply demonstrated that coping is a culturally-contingent process, and occurs differently within different social and cultural contexts (Rich et al., 1995; Oliver-Smith 1996; Gist and Lubin 1999). Different societies have culturally specific ways of managing trauma, and this needs to be recognized at the onset of post-disaster aid and reconstruction efforts. Extending from the literature, we argue that recovery processes are also historically- and materially-contingent, as they are part of broader cultural and social trajectories, and carried out in meaningfully constituted environments which are integral to their enactment. This is an important point that we expand upon below.

From a cultural heritage perspective, we are very sympathetic to the issue of continuity, and argue that one of the most important aspects of recovery in the immediate post-trauma period is re-establishing familiarity. This common-sense argument is supported by Omer's 'continuity principle' which 'stipulates that through all stages of disaster, management and treatment should aim at preserving and restoring functional, historical, and interpersonal continuities, at the individual, family, organizational and community levels' (Omer and Alon 1994, p. 274). We agree with this basic premise, and feel that re-engaging with pre-existing social and cultural contexts is fundamental to community recovery, an argument that is also supported by a number of other authors (Omer and Alon 1994; de Vries 1995; Gist and Lubin 1999, etc.).[9] From this we see the ultimate benchmark of the success of recovery efforts as how well communities are able to continue as cohesive social and cultural entities in the aftermath of reconstruction. Given the scale of relief and reconstruction efforts in many post-disaster situations and the increasing internationalization of involvement, this question is assuming more and more significance.

VERNACULAR LANDSCAPES, SOCIAL PRACTICES AND RECOVERY

Existential space is a constant of production and reproduction through the movements and activities of members of a group. It is a mobile rather than a passive space for experience. It is experienced and created through life-activity, a sacred, symbolic and mythic space replete with social meanings wrapped around buildings, objects and features of the local topography, providing reference points and planes of emotional orientation for human attachment and involvement. (Tilley 1994, p. 16).

...losing access to places of cultural and social significance, and the resulting loss of connections to people, undermines the community's ability to turn its 'wheels of healing (de Vries 1995, p. 379).

So far we have made the argument that recovery needs to be conceptualized within social and cultural frameworks in order to more fully understand the complexities of post-trauma community coping processes. We propose that programmes that are sensitive to the cultural and social dimensions of recovery have a higher likelihood of sustainability and success. In this section we make a direct connection between these social and cultural processes and vernacular landscapes.[10] This is especially important because the bulk of what is typically referred to as 'reconstruction' directly involves the built environment, and in the case of Aceh modifying it as part of 'building back better'. There are two major issues that need to be considered when restructuring peoples' environments after a disaster. First, landscapes are culturally understood, meaningfully constituted, and usually the result of long-term accumulative processes. They are places filled with cultural significance that ground communities. Second, the material world is integral to the enactment of most social practices. People need to have the appropriate cultural and physical settings to best carry out practices that are fundamental to social reproduction, and processes potentially relevant for reconstruction and recovery. Building back differently is not only potentially disorientating to communities looking to re-establish connections with familiar physical settings because things look, feel, and seem foreign, but also because many of the latent coping and recovery mechanisms that communities need to draw upon in such times are interrelated with the material world in which they existed. Even in the most heavily-damaged landscape, a conceptual 'familiar' gives a solid framework for community reconstruction processes, providing understood targets, tangible benchmarks of success, and reassurance.

The work of the cultural anthropologist Oliver-Smith has usefully emphasized this relationship between recovery and place.

Recent research emphasizes the importance of place in the construction of individual and community identities, in the encoding and contextualization of time and history, and in the politics of interpersonal, community, and intercultural relations. Such place attachments mean that the loss or removal of a community from its ground by disaster may be profoundly traumatic. (Oliver-Smith 1996, p. 308)

We argue that at least in some cases, 'building back better' undermines the functionality, vitality, and cultural importance of local built environments and implicit social mechanisms that are important for both long-term social recovery and comprehensive community participation within relief and reconstruction efforts. As we show, many of the basic 'assets' that were damaged in Aceh by the Tsunami and need to be 'reconstructed' are deeply involved in the enactment of social practices, play powerful roles in people's attachment to places, senses of identity, and are benchmarks for some semblance of 'normalcy'. It is common within externally-driven reconstruction and development programmes that core components within local cultural landscapes are viewed in distant and practical terms, and the deeper importance of such features, spaces and places to the constitution of communities is neglected. This needs to be redressed. We now turn to the results from some of our fieldwork to further explore such issues within the reconstruction of post-Tsunami Aceh. We use a few examples to show the connections between elements of the built environment and social practices important to community recovery.

SURVEY OF CULTURAL HERITAGE AND RECONSTRUCTION

We have been present in Aceh since the Tsunami, working in various capacities with NGOs, academics, as well as jointly coordinating surveys with the Aceh Heritage Community.[11] We have conducted six months of detailed village-level field survey since early 2005 in Tsunami-affected areas focusing upon issues of cultural heritage and reconstituting society. We collected data during three major surveys of over 150 sites in and around Banda Aceh designated to be culturally and/or historically important based on a pre-Tsunami inventory of 'heritage sites' held by the Aceh museum, adjusted by our team to include a wider spectrum of non-monumental and colonial sites. These surveys have been carried out at annual intervals since the Tsunami, allowing us to observe the relationship between reconstruction processes and cultural heritage sites. This serves the practical role of assisting local authorities and international organizations to manage heritage sites as part of post-disaster reconstruction efforts. The data also allow us to test our hypothesis that heritage sites are important for community recovery because they serve as tangible anchors that help communities reorient themselves.

In our fieldwork we talked extensively with inhabitants familiar with local geography to locate sites. Through many informal discussions, which often entailed explaining what we were looking for, and why, we built up a much clearer image of what local inhabitants thought were important components of the built environment that fell outside our more formalized understanding of heritage. People could easily identify sites that were most meaningful to them, and it was often clear that the context was heavily influenced by reconstruction. We found that there are a number of types of structures and places that people in Aceh identify as important within localized conceptions of culture and heritage that have some relevance for community reconstruction.

To gather more systematic qualitative data on the role that this 'vernacular' heritage plays within processes of community recovery, we conducted a detailed field survey in February 2007 in which over 250 respondents were interviewed (the average length of interview was one hour). We visited 12 villages in both affected areas, and outside the damage zone, the latter to establish control variables. The sites are based within an hour of Banda Aceh, which was one of the most heavily-damaged areas, and also the epicentre of the aid efforts. The survey was carried out by teams from the Aceh Heritage Community accompanied by two staff members from the Aceh Museum and the two authors. We held a methodology training session for all staff involved in the survey, and the fieldwork was conducted under the constant field supervision of both authors who co-directed the project. With the exception of Daly all the interviewers are Acehnese, mostly university undergraduates who have backgrounds in architecture, with an interest in cultural heritage. The interviews were conducted in either Bahasa Indonesia or Acehnese, depending on the preferred language of the respondents. All field notes were made in Indonesian, and all recorded interviews transcribed into Indonesian for analysis.

During the course of the interviews, we carried out detailed discussions with respondents about what they identified as important components of vernacular heritage within their communities, and how such places were important to processes of recovery and reconstruction. The main goal of the questions in the survey was to identify the material components that are culturally meaningful within the construction and maintenance of cultural identities and social practices, and which served practical roles in areas pertinent to reconstruction, such as establishing the contexts necessary for meaningful community dialogue.

The survey provided a wealth of information about the elements of vernacular cultural heritage that are recognized as important to village inhabitants. While the results showed scope for localized variability, many of the *gampongs*[12] shared a similar orientation and possessed the same categories of elements that played clearly-defined roles. Our discussion of the results is framed by three scales of materiality: the village as an entity, structures that service broader community needs, and individual family dwellings. This is not meant to be a comprehensive accounting of all elements of vernacular heritage within Acehnese *gampongs*, but rather to draw on a few examples to illustrate the points raised above.

DISCUSSION OF SURVEY RESULTS

Our data fully support that *gampongs* are not just administrative categories or purely physical entities, but rather well-established and functioning social mechanisms firmly grounded in a blend of cultural and religious traditions. This has been mentioned in a few studies of reconstruction in Aceh (Mahdi this volume), and is commonly understood by the Acehnese and foreigners who have invested significant time in Aceh. The cohesiveness of *gampongs* was brought up in many of our interviews, which emphasized the communal nature of social practices and the deeply-embedded social order of the *gampong*. Many facets of society in Aceh are heavily structured at the *gampong* level, with prescribed hierarchies, leadership, manners of public debate and discussion, and formal decision making, all of which are essential for the realization of 'participatory' reconstruction and development practices. In almost all our discussions, respondents' replies acknowledged this collective. It came across that many of the tools needed for organizing and driving local reconstruction efforts were inherent within *gampongs*, with most of what was lacking being material and financial resources.

It also was made clear that the vital roles that *gampongs* play within Acehnese identity, structuring social networks, decision making, etc. are embodied within the vernacular landscapes of Acehnese villages. The responses show that there are components within *gampong* layouts that are important for the enactment of a wide range of social practices and engagements. People recognize and discuss a *gampong* as a tangible entity, or a collection of such. For example, any discussion about community debate and decision making typically involves the village head (*geuchik*), an assortment of village elders, and occurs within specific places, such as mosques (elaborated upon below) and coffee shops. The fact that it is difficult to find a natural separation between physical elements, people and processes speaks loudly about the interconnectedness of all three. Our data supports that the physicality of *gampongs* serves to reinforce their social efficacy.

Two of the most important examples of this are the mosque and *meunasah*,[13] which are both standard parts of most Acehnese *gampongs*. Most respondents identified these features as the most important elements of 'cultural heritage' in each village, which occurred irrespective of the actual age of the structures. When posed with clarification questions, respondents steadfastly insisted that these were items of Acehnese *heritage*, even in the cases of mosques that were built within the last decade, and in some cases clearly by or with the support of foreign agents.[14] Respondents pointed out that mosques and *meunasahs* play a fundamental role at a number of different levels, as venues for formalized religious and ritual observances, and as spaces for social interaction and discussion. Most people we spoke with about this were not able to conceptualize an Acehnese community in the absence of these features, which further emphasizes their importance.

Respondents discussed the roles of both types of structures within post-disaster recovery and reconstruction efforts. When talking about community-initiated responses, respondents pointedly discussed how such structures were missed, and should have been amongst the first places concentrated on during reconstruction, as it was from these that the rest of the community extends. Furthermore, the absence of such places made it difficult for communities to conduct discussions concerning reconstruction on familiar terms, and placed interaction with external agents outside culturally-suitable confines. This was mentioned by a number of respondents, especially when discussing life in temporary barracks where much of the community 'consultation' by NGOs occurred, and which were often lacking much of the physical and cultural infrastructure of Acehnese *gampongs*.

Interestingly, mosques have taken on an added symbolic importance as related to both the reasons for the disaster — which many Acehnese attribute to Allah's dissatisfaction, and the source of hope for survival and recovery — the strength of faith and Islam.[15] A number of mosques have become iconic due to their better rates of survival compared with other structures in Tsunami-hit areas, with some, such as the mosque at Lampuuk, attaining international fame.[16] When interviewing residents around Lampuuk, it was very common for respondents to stress that the mosque survived whereas everything else was destroyed, attributing this to divine will rather than structural soundness, and that the mosque was a source of strength for people sifting through the wreckage of their lives and communities. Stories about almost dying, but seeing the mosque and persevering have become part of local folklore. Regardless of the veracity of specific accounts along these lines, it is clear that the survival of such structures has become a firmly-established part of local narratives of faith and perseverance, and they seem to be physical anchors around which both physical and social recovery has coalesced. This is an important example of people relying upon culturally-meaningful parts of their environments to provide emotional and psychological buoyancy. This recourse is clearly not possible outside the confines of a familiar, even if heavily damaged, setting, and supports the idea of not removing inhabitants for too long from their land.

The reconstruction of housing has become emblematic of the post-Tsunami response efforts in Aceh. As housing is both vital for basic human needs and the most intimate material setting for individual and family engagements, it is right to place significant emphasis upon it. However, throughout Aceh, the majority of the housing (re)construction has been driven by NGOs and filtered through a complicated arrangement of construction companies and subcontractors. This has resulted in delays, confusion, and, most importantly, the construction of housing that is often inappropriate within local cultural and social contexts. Furthermore, as many of our respondents pointed out, all phases of the construction were largely carried out by non-locals.[17]

Respondents in many areas were quick to mention the physical problems of the houses, which ranged from leaky roofs, to the lack of kitchens! A number of respondents also mentioned how life was different now because the villages and houses were different.[18] Several respondents talked in detail about the gendered division of spaces within homes, and the various implications of the spatial arrangements of housing within *gampongs*. It was brought up that whereas mosques and *meunasahs* are spaces for male interaction and discussion, domestic settings are critical for female social interaction and dialogue. Traditionally, the areas under the raised Acehnese houses and around the kitchen when the houses were not raised are keys spaces for women to gather and talk about community issues. This, coupled with the spatial arrangements of houses within new village layouts, has left women in some villages feeling more isolated. This is something that we witnessed repeatedly during the survey, with women spread out, usually in close proximity to their homes, and by themselves or only in small groups. This contrasted with the natural congregation of males at the coffee shops.[19]

One of the villages that we visited, Kampung Jawa, is of especial interest to this discussion as one of the main providers of 'shelter' there, Muslim Aid, made a conscious effort to offer a style of housing that fit within Acehnese traditions. Inhabitants within the village were able to choose between a 'modern' concrete house, built on the ground level, or a raised house modeled on the *rumah Aceh*, the traditional style of Acehnese homes. There was no obvious consensus on style, which is interesting in its own right, but from our interviews many of the respondents, regardless of which style they chose, were happy that Muslim Aid offered a traditional style of housing — whether they took it or not. People saw this as a respectful gesture, and a clear case in which a reconstruction project was sensitive to local wishes and cultural habits. However, within a year of construction, just about all the respondents complained that while the idea was a good one, the material was of poor quality, they were worried about the use of asbestos,[20] and people were 'kept awake at night by the sound of the termites chewing wood'. Inhabitants grew increasingly frustrated by the fact that the NGO subcontracted the work out and the end result was unsatisfactory housing. It is a great shame that such an initiative failed due to technical issues, as it is one of the few cases that we encountered in Aceh of an NGO specifically replicating traditional housing models.

Overall, housing is an issue that has contributed significant obstacles to recovery, including forcing displaced people to react to very different and unfamiliar spatial parameters which are not sympathetic to pre-existing conditions, both in the temporary barracks, and within the newly constructed housing.[21] A number of the respondents had just moved back to their villages after having spent almost two years in temporary facilities far removed from their land. People openly questioned why they were moved and kept off their lands for so long while waiting for other people to clear the debris and build housing for them. Furthermore, some of the respondents mentioned that during the period of displacement, the remains of communities were sometimes scattered, and the circumstances that they found themselves in often did not match familiar social conditions.[22] This highlights some practical issues, but also suggests connections with *gampongs* and a desire in at least some to return to familiar settings as quickly as possible.

The problems of providing housing have spilled over into related issues of land rights and entitlement, which respondents mentioned in a number of ways. They talked in detail about the mix of different people within reconstructed villages, and the presence of 'outsiders' taking up residence within the 'tsunami houses'.[23] In some cases these were relatives of victims who moved in to claim family land, but it was very common to hear about Javanese workers, or people from elsewhere in Aceh settling in reconstructed villages because of proximity to employment opportunities.

Finally, the basic logic behind large-scale house construction by NGOs is predicated upon concepts of land tenure, titling and ownership that are not consistent with pre-existing practices, in which land ownership was understood within the framework of *adat*, or local cultural tradition. While the RALAS[24] project has worked extensively with local communities to reconcile this through a rigorous programme of community mapping and titling, the entire process of house allocation is fraught with complications, and has become wide open for abuse (Fitzpatrick, this volume). This stems from the failure of NGOs to see housing within a broader Acehnese context of community spatial arrangements and land ownership.

A final point of contention brought up by some of our respondents is the jealously that has sprung up between — and in some cases even within — *gampongs* because of the differential allocation of resources for housing. Dozens of foreign organizations have been involved in house construction along hundreds of miles of coastal areas, with only limited oversight and coordination. The variety of different blueprints, materials used, and amenities included have created an unequal landscape in which random factors often led to certain areas receiving certain kinds of support. This was discussed by respondents in a number of villages who are acutely aware that some villages have got 'better' provisions, and could be a potential source of long-term tension.

CONCLUDING REMARKS

While there is value in studying the impact of disaster and reconstruction upon 'heritage sites' for the benefit of international heritage management practices, it is tremendously important within the context of post-disaster reconstruction to better understand the constellation of meaningfully constituted places that forms the vernacular cultural heritage within communities. Such sites and places are part of the fabric of everyday social life, and elements of the landscape that people gravitate towards in times of trauma and disorientation. As described above, people in Aceh were very clear about the sites and places that they found significant for enabling and contextualizing their individual- and community-level responses, and allowing them access to their pre-Tsunami lives.

Second, many sites within the vernacular landscape actively serve important roles within the day-to-day enactment of important social functions, such as providing spaces for community deliberation, conflict resolution, and decision making, all of which are fundamental for recovery. Many of the processes through which meaningful discussions about the past, present and future of traumatized people are facilitated not by foreign intervention, but rather by access of the population to the basic culturally-appropriate venues and contexts in which different forms of discussion occur. This not only involves community leadership, social networks and hierarchies, but also the material settings within which communities know how to interact.

In the absence of all the above, meaningful and comprehensive community participation is not possible.

Interestingly, it was immediately clear that within Acehnese understandings of vernacular cultural heritage there is a tremendous conflation of the cultural, the historical and the religious. This is not a surprise given the level of religious observance within Aceh which has both become more formalized by the implementation of sharia, and reinforced by a religious revival as part of the overall response to the trauma (Miller 2010). However, it is important to consider this within the context of understanding how cultural and religious sites and places serve as venues for a wide range of social functions ranging from everyday interaction to special ceremonies. Our discussions with respondents show that such sites play a simultaneous role as integral to religious identities, and as spaces where discussions necessary for all phases of relief, reconstruction and development are carried out. This was often an issue for international organizations which either have specific rules against contributing towards religious structures, and/or are staffed by people with a limited real appreciation for the role of faith within communities.[25]

Our experiences in Aceh show that there is great value in carefully considering pre-existing social and cultural conditions, and appreciating that these are part of the complexities of community-level coping and recovery processes. The prevailing 'build back better' attitude on behalf of the reconstruction and development industries, coupled with significant levels of disconnect between the givers and receivers of aid and ignorance of Acehnese cultural and social practices has undermined important mechanisms for internally-driven social rehabilitation, and effective and useful distribution of aid resources. Clearly, there is much room for outside agents to assist in processes of recovery, but as we argue vigorously, any such efforts that impede exercising latent coping mechanisms, or a community's creation of new methods for responding to extraordinary circumstances can be detrimental, and contradict the basic logic behind relief and reconstruction interventions. Steering away from culturally-familiar settings and practices not only provides further disruption and disorientation, but also poses serious obstacles that make it difficult for communities to find their own intuitive paths to recovery. We caution that explicit external agendas in which relief and aid is contingent upon or targets social transformation can contribute towards further disorientation, and loss of involvement in key phases of recovery.[26]

> ...the indifference towards local wisdom will bury the home and dream of a new Aceh from the first ground broken for reconstruction (YAKKUM Emergency Unit website <http://www.yeu.or.id/about_us.php>).

Simply put, when operating in foreign environments it should be conventional practice to be familiar with and respect local wisdom. Unfortunately, this was not the standard operating procedure for the reconstruction work in Aceh in spite of the rhetoric. In the urgency for bureaucrats and aid agencies to obtain tangible statistics of success to parade before their constituents and donors, to placate the population within Aceh, and to achieve their mandate of 'building back better', there has been widespread and systematic neglect of local wisdom. Whether born from ignorance or arrogance, the vast majority of the NGOs and governmental organizations effectively pursued community-level reconstruction from a top-down perspective, and allowed much of the process to be determined by contractors and outside consultants. Furthermore, the processes of community consultation by NGOs and other development agencies during the main period of needs assessment and reconstruction planning were deeply flawed. Our research makes it clear that processes are contingent upon sets of both physical institutional infrastructure and social apparatuses that create *gampong* identity and support community discourse. The wide-scale absence of both effectively reduced — if not crippled — many of the efforts to 'involve' local community leadership and members in reconstruction planning. Realizing the ideal of 'participatory' reconstruction and development practices cannot be achieved without understanding the local cultural landscape and appreciating the social nuances embodied within it. It is of the outmost importance to understanding how reconstructing the built environment could have been better synchronized with easing social trauma, and allowing the inhabitants of these regions to have the maximum (albeit under heavily-distressed circumstances) opportunity to re-orientate themselves in a shattered landscape.

Notes

1. Estimates have put the total post-tsunami expenditure at over 12 billion USD while the 'Master Plan for Rehabilitation and Reconstruction for the Regions and People of the Province of Nanggroe Aceh Darussalam and Nias Islands of the Province of North Sumatra', the main and official guide for rehabilitation and reconstruction from the Indonesian government, puts the proposed funding needed for rehabilitation and reconstruction submitted by ministries/institutions until 2009 at Rp58.3 trillion. The funds to support this comes from 1) the Indonesian government which has pledged Rp5.9 trillion (including the Rp3.9 trillion moratorium from Paris Club but apart from funds coming from Departments and Institutions within NAD province and Nias in the form of decentralization funds, assistance duties, and central institutions funds, the

judicial sector and the financial sector) and 2) foreign grants of Rp15.7 trillion from bilateral sources and Rp7.7 trillion from multilateral sources. There was also a US$300 million Asia Development Bank grant. The estimated total funds pledged by the private and public sectors amounts to Rp13.5 trillion. The Joint Evaluation of International Response to the Indian Ocean Tsunami synthesis report put the total of international flows of funding at US$13,503 million (Telford & Cosgrave 2006, p. 81).

2. This can be seen very acutely in the heated debate that has slowed down the reconstruction of the World Trade Center site following the September 11th, 2001 attacks (Vale & Campanella 2005). This case highlights the complicated and often contentious competing voices vying to shape post-disaster reconstruction.

3. A perusal of the websites as well as documents of major donor organizations and implementing agencies show the adoption of the "build back better" framework by organizations such as the UNDP, UNICEF, World Bank, World Vision, the Human Rights Watch, and the BRR, to name just a few.

4. At this point it is important to make a clear statement of support for the vast majority of talented and well meaning individuals and organizations (both local and international) who worked extremely hard to improve the situation in Aceh and other tsunami affected areas. In many ways a lot of our findings resonate with the complaints and frustrations experienced and expressed by many of the aid workers who have committed huge amounts of time and energy to the reconstruction processes.

5. There are a number of sources on historical and cultural information about Aceh, for example: Snouck 1906; Bowden 1991, 1993*a*, 1993*b*; Reid 1969, 2004, 2006; Feener et al. 2011, etc.

6. The broad political and social implications touched on here will be explored fully in future publications.

7. One ready example is the massive emphasis placed upon 'gender' programs. None of the major aid and reconstruction organizations would have had the kind of access that they were granted if they had come specifically to recast gender roles within Acehnese society. This has the potential to confuse very different issues: aiding a society recovery *and* instigating major social transformation. Furthermore, such attempts usually lack a sophisticated understanding of pre-existing gender roles in Aceh, and frames things within overt stereotypes of how Muslim and Indonesian — read developing world — women are treated. Discussions with people in Aceh already suggest that even people sympathetic with the broader ideas of changing the gender dynamics in Aceh are weary of all of the focus on gender.

8. In a fascinating discussion of urban destruction and reconstruction Vale & Campanella build a well supported case that it is the historical exception for heavily damaged cities NOT to recover, drawing almost exclusively upon case

studies that pre-date the recent internationalization of relief and reconstruction processes (2005).

9. There is extensive literature focusing upon how disasters and subsequent responses expose or even reinforce pre-existing social inequalities (see for a summary Gist and Lubin 1999, p. 49). In recent years this has become a major issue for donors and aid agencies who are rightly cautious about serving the interests of structures that further disadvantage segments of society on the basis of gender, class, age, race, political affiliation, etc. While perhaps laudable at one level, this is used as a justification for programs that intentionally work outside of culturally understood contexts. This raises critical questions about passing moral judgments in the absence of clear moral authority, and determining the basic aim of relief and reconstruction processes. Given that some matter of injustices and inequalities exist in all societies, it is strange and perhaps hypocritical for donor countries to make such distinctions: a point that is not lost on most locals within reconstruction and development situations. More importantly, we believe that decision-making needs to be confined within a strict framework of aiding recovery in traumatized communities, and initiating social transformation as part of aid interventions should be the exception rather than the rule for external organizations. We are inclined to agree with Vale and Campanella's statement that "recovery must also entail some sort of return to normalcy in the human terms of social and economic relations, even if that so called normalcy merely replicates and extends the inequities of the pre-disaster past" (Vale & Campanella 2005), and accept the consequences of pre-existing inequities persisting. We remain unconvinced by the long-term success rate of programs that insist on meddling with existing social realities. Furthermore, there is support in the literature that disasters can bring about opportunities for the entrance of new groups into community power structures that is the result of local dynamics (Aronoff & Gunter 1992; Bolin & Stanford 1989; Couto 1989; Gibbs 1982; Oliver-Smith 1996; Rich et al. 1995). If real change is going to occur, it is preferable that local inhabitants have authorship of it, ownership of their successes, and responsibility for their failures.

10. We are referring to the everyday lived-in environment, which includes human constructed features, as well as places and spaces that are culturally meaningful. There is extensive empirically grounded research within human geography, anthropology, archaeology and architecture dealing with meaningfully constituted spaces and places (For some relevant examples refer to Gregory & Urry 1985; Hough 1990; Gupta & Ferguson 1992; Bender 1993; Ingold 1993; Crumley 1994; Barrett 1994; Hirsch & O'Hanlon 1995; Appleton 1996; Feld & Basso 1996; Bradley 1998; Ashmore & Knapp 1999; Joyce & Gillespie 2000; Ashmore 2002; King 2003; Low & Lawrence-Zúñiga 2003; Forbes 2007).

11. The Aceh Heritage Foundation, founded by one of the authors, works to preserve cultural heritage in Aceh and to raise awareness about the history of Aceh.

12. *Gampung* is the Acehnese word for village, but as discussed below, it is an important concept that far transcends a generic definition of village.

13. A meunasah is a combination of prayer room and community space. Its importance is summed up in the following statement: "the concept of the meunasah in Aceh's societal structure is that of a village mushalla. And yet, a meunasah is not just a place for worshipping. It also fulfills the function of community… almost every aspect of village life in Aceh is centered in the meunasah. All kinds of cultural products grew out of the meunasah…" (*YAKKUM Emergency Unit* website <http://www.yeu.or.id/about_us.php>) The meunasah plays a critical role as an intermediary between locals and wider levels of government and jurisprudence. For example, it is common for legal cases to be less formally resolved within the meunasah, avoiding the much more complicated and expensive process of taking matters to high courts. This is just one of the roles it plays to mitigate social tension (M. Feener *pers com*).

14. It is common to have mosques, even prior to the tsunami, which were built with outside support. In particular, funds from Persian Gulf states have contributed prominently to the construction of religious structures. It is interesting to note that there is a long history of external influence on Mosque construction in Aceh, including Ottoman, and far less obviously Dutch efforts. Yet, regardless of the source of funding, or architectural blueprints, mosques are seen as being Acehnese, seemingly following the rather faulty logic of mosques are Muslim, Acehnese are Muslim, therefore the mosques are Acehnese!

15. There is an established body of literature discussing faith and piety following disasters. It is a common phenomenon for people to attribute natural disasters to higher powers, and for communities to seek solace within faith (Bushnell 1969; Pargament & Hahn 1986; Ahler & Tamney 1964; Bradfield, Wylie & Echterling 1989; Smith 1978; Gist and Lubin 1999; Oliver-Smith 1996; etc.).

16. The mosque at Lampuuk became a symbol of the strength of Allah all throughout the Muslim world, greatly aided by the staggering aerial photos taken immediately after the tsunami showing the mosque standing alone amidst vast devastation. In the months following the tsunami, Lampuuk became a common destination for visiting delegations, politicians and even tourists, including former US presidents Bush and Clinton. Informal discussion during the course of our field work suggests that the presence of such sites, and the 'prestige' associated with them, has influenced the geography of aid distribution, with governments and organizations gravitating in some cases towards the better 'photo opportunities.'

17. It was common practice to import both specialists and manual laborers for large-scale reconstruction processes. This further ensured that both local input and involvement was limited. An interesting study on housing following the earthquake in Gujarat demonstrated that on a number of levels the most effective solution to house reconstruction was to simply provide material and financial assistance, and allow local inhabitants to deal with the design and construction

by themselves. The study shows that this approach was more efficient, cost-effective, and there were far higher levels of satisfaction with the end products than in Aceh (Barenstein 2006). It is a shame that such 'user-driven' models were not more widely employed in Aceh for a number of reasons, and it is much more likely that such an approach would have contributed towards far more culturally suitable homes being built, and created a focal point for direct involvement: people literally rebuilding their homes and communities.

18. Unfortunately our data did not get to a level of nuance needed to really tell the full story with regards to housing, as there is both a tremendous variation of housing styles built by many organizations, and set limits based upon the methodology employed. Therefore we are not able to provide a more sophisticated model here.

19. Coffee shops play very important roles within the social life of males in Aceh, and this is one feature that we found reconstructed in some form in all villages we visited. This is just one of a number of other features within Acehnese landscapes that registered of interest within our study, but we lack the space to elaborate on all elements of the built environment in this chapter.

20. This is one of the prevalent rumors going around the village, but we were not able to confirm that there was actually asbestos within the houses.

21. However, there are examples in which people seem happy with 'new' housing that is very different from their pervious homes. In particular, many residents of Lampuuk village, which was reconstructed by the Turkish Red Crescent, reported that they were satisfied with their new 'modern' housing. While this is most likely related to how well constructed and furnished the houses were, respondents also noted that they had felt very involved in the reconstruction process in their village. This supports the idea mentioned above that there are differences when inhabitants feel that they are behind the changes brought about, rather than the recipients of externally imposed change.

22. This closely echoes Gist and Lubin's point that "Following large-scale disasters, people in charge of relocation efforts may, out of ignorance or simple expediency, disregard natural groupings traditionally existing within communities, and many victims must rely on temporary housing that seldom reflects pre-disaster personal relationships and neighborhood patterns" (Gist and Lubin 1999, p. 41).

23. We had to begin all interviews by establishing if the potential respondent was originally from the *gampung* in question. We were initially very surprised by the number of people who had not lived in these areas until after the tsunami.

24. See Fitzpatrick 2008*a*, 2008*b*, 2008*c*, & 2008*d* for more detailed information on the RALAS programme, as well as Fitzpatrick in this volume.

25. This is typical of the skepticism about local religious and cultural beliefs and practices that is widespread within the ranks of international NGO workers in Aceh, and other aid situations.

26. This is a much broader critique of the relief, reconstruction and development processes than can be dealt with in this chapter. A more comprehensive discussion

will be presented within a monograph on post-disaster social transformation (Daly Forthcoming).

References

Ahler, J. G. and Tamney, J. B. "Some Functions of Religious Rituals in a Catastrophe". *Sociological Analysis* 25 (1964): 212–30.

Appleton, J. *The Experience of Landscape.* New York: John Wiley, 1996.

Aronoff, M. and Gunter, V. "Defining Disaster: Local Constructions for Recovery in the Aftermath of Chemical Contamination". *Social Problems* 39 (1992): 345–65.

Ashmore, W. "Decisions and Dispositions: Socializing Spatial Archaeology". *American Anthropologist* 104, no. 4 (2002): 1172–83.

Ashmore, W. and Knapp, A. B. eds. *Archaeologies of Landscape: Contemporary Perspectives.* Cambridge: Blackwell, 1999.

Barenstein, J. "A Comparative Study of Six Housing Reconstruction Approaches in Post-Earthquake Gujarat". HPN Network Paper 54 (Humanitarian Practice Network, March 2006).

Barrett, J. *Fragments from Antiquity: An Archaeology of Social Life 2900–1200 B.C.* Oxford: Blackwell, 1994.

Bender, B. ed. *Landscape: Politics and Perspectives.* Oxford: Berg, 1993.

Bennett, J., Bertrand, W., Harkin, C., Samarasinghe, S. and Wikramatillake, H. *Coordination of International Humanitarian Assistance in Tsunami-Affected Countries.* London: Tsunami Evaluation Coalition, 2006.

Bolin, R. and Stanford, L. "The Northridge Earthquake: Community-Based Approach to Unmet Recovery Needs". *Disasters* 22, no. 1 (1998): 21–38.

Bonanno, G. "Trauma, and Human Resilience: Have We Underestimated the Human Capacity to Thrive After Extremely Aversive Events?". *American Psychologist* 59, no. 1 (2004): 20–28.

Bonanno, G. and Keltner, D. "Facial Expressions of Emotion and the Course of Conjugal Bereavement". *Journal of Abnormal Psychology* 106 (1997): 126–37.

Bowen, John Richard. *Sumatran Politics and Poetics: Gayo History, 1900–1989.* New Haven: Yale University Press, 1991.

———. "Return to Sender: A Muslim Discourse of Sorcery in a Relatively Egalitarian Society, the Gayo of Northern Sumatra". In *Understanding Witchcraft and Sorcery in Southeast Asia*, edited by C. W. Watson and Roy Ellen. Honolulu, Hawaii: University of Hawaii Press, 1993*a*.

———. *Muslims Through Discourse.* Princeton: Princeton University Press, 1993*b*.

Bradfield, C., Wylie, M. L. and Echterling, L. G. "After the Flood: The Response of Ministers to a Natural Disaster". *Sociological Analysis* 49 (1989): 397–407.

Bradley, R. *The Significance of Monuments: On the Shaping of Human Experience in Neolithic and Bronze Age Europe.* London: Routledge, 1998.

Brickman, P., Rabinowitz, V., Karuza, U., Coates, D., Cohn, E. and Kidder, L. "Models of Helping and Coping". *American Psychologist* 37, no. 4 (1982): 368–84.

Bushnell, J. H. "Hupa Reaction to the Trinity River Floods: Post-hoc Recourse to Aboriginal Belief". *Anthropological Quarterly* 42 (1969): 316–24.

Cardena, E., Holen, A., McFarlane, A., Solomon, Z., Wilkinson, C. and Spiegel, D. "A Multisite Study of Acute Stress Reactions to a Disaster". In *DSM-IV Sourcebook*, edited by T. A. Widiger, A. J. Frances, H. A. Pincus, R. Ross, M. B. First, W. Davis and M. Kline. American Psychiatric Association, 1994, pp. 377–91.

Christoplos, I. *Links between Relief, Rehabilitation and Development in the Tsunami Response*. London: Tsunami Evaluation Coalition, 2006.

Crumley, C. ed. *Historical Ecology: Cultural Knowledge and Changing Landscapes*. Santa Fe, New Mexico: School of American Research Press, 1994.

Cuoto, R. A. "Catastrophe and Community Empowerment: The Group Formulation of Aberfan's Survivors". *Journal of Community Psychology* 17 (1989): 236–48.

Daly, P. *Zero Hour: Social Transformations and Community Recovery in Post-Tsunami Aceh*. Forthcoming.

De Vries, M. W. "Culture, Community and Catastrophe: Issues in Understanding Communities under Difficult Conditions". In *Extreme Stress and Communities: Impact and Intervention*, edited by S. E. Hobfoll and M. W. de Vries, Dordrecht. The Netherlands, 1995, pp. 375–93.

Feener, M., Daly, P. and Reid, A. eds. *Mapping the Acehnese Past*. Leiden: KITLV Press.

Feld, S. and Basso, K. H. eds. *Senses of Place*. Santa Fe, New Mexico: School of American Research, 1996.

Fitzpatrick, D. "Managing Conflict and Sustaining Recovery: Land Administration Reform in Tsunami-Affected Aceh". ARI Working Paper 004 (Singapore: Asia Research Institute, National University of Singapore, 2008*a*).

—. "Women's Rights to Land and Housing in Tsunami-Affected Aceh, Indonesia". ARI Working Paper 003 (Singapore: Asia Research Institute, National University of Singapore, 2008*b*).

—. "Access to Housing for Renters and Squatters in Tsunami-Affected Aceh, Indonesia". ARI Working Paper 002 (Singapore: Asia Research Institute, National University of Singapore, 2008*c*).

—. "Housing for the Landless: Resettlement in Tsunami-Affected Aceh, Indonesia". ARI Working Paper 001 (Singapore: Asia Research Institute, National University of Singapore, 2008*d*).

Forbes, H. *Meaning and Identity in a Greek Landscape: An Archaeological Ethnography*. Cambridge: Cambridge University Press, 2007.

Gibbs, L. "Community Response to an Emergency Situation: Psychological Destruction and the Love Canal". *American Journal of Community Psychology* 11 (1982): 116–25.

Gilbert, D. and Silvera, D. "Overhelping". *Journal of Personality and Social Psychology* 70, no. 4 (1996): 678–90.

Gist, R. and Lubin, B. *Response to Disaster: Psychological, Community, and Ecological Approaches.* London: Taylor and Francis, 1999.

Gregory, D. and Urry, J. eds. *Social Relations and Spatial Structures.* Basingstoke: MacMillan, 1985.

Gupta, A. and Ferguson, J. "Beyond 'Culture': Space, Identity, and the Politics of Difference". *Cultural Anthropology* 7, no. 1 (1992): 6–23.

Hirsch, E. and O'Hanlon, M. eds. *The Anthropology of Landscape: Perspectives on Place and Space.* Oxford: Clarendon Press, 1995.

Hough, M. *Out of Place: Restoring Identity to the Regional Landscape.* New Haven, CT: Yale University Press, 1990.

Ingold, T. "The Temporality of the Landscape". *World Archaeology* 25, no. 2 (1993): 152–74.

Joyce, R. A. and Gillespie, S. D. eds. *Beyond Kinship: Social and Material Reproduction in House Societies.* Philadelphia: University of Pennsylvania Press, 2000.

King, T. F. *Places That Count: Traditional Cultural Properties in Cultural Resource Management.* Walnut Creek, CA: Alta Mira, 2003.

Klein, N. *The Shock Doctrine: The Rise of Disaster Capitalism.* New York: Metropolitan Books, 2007.

Low, S. M. and Lawrence-Zúñiga, D. eds. *The Anthropology of Space and Place: Locating Culture.* Cambridge: Blackwell, 2003.

Miller, M. A. "The Role of Islamic Law (Shari'a) in Post-Tsunami Reconstruction". In *Post-Disaster Reconstruction: Lessons from Aech,* edited by M. Clarke, I. Fanany and S. Kenny. Earthscan.

Norris, F. and Kaniasty, K. "Received and Perceived Social Support in Times of Stress: A Test of the Social Support Deterioration Deterrence Model". *Journal of Personality, Psychology and Social Psychology* 71, no. 3 (1996): 498–511.

Oliver-Smith, A. "Anthropological Research on Hazards and Disasters". *Annual Review of Anthropology* 25 (1996): 303–28.

Omer, H. and Alon, N. "The Continuity Principle: A Unified Approach to Disaster and Trauma". *American Journal of Community Psychology* 22, no. 2 (1994): 273–87.

Pargament, K. I. and Hahn, J. "God and the Just World: Causal and Coping Attributions to God in Health Situations". *Journal for the Scientific Study of Religion* 25 (1986): 193–207.

Reid, A. ed. *Verandah of Violence: The Historical Background of the Aceh Problem.* Singapore: Singapore University Press, 2006.

———. *An Indonesian Frontier: Acehnese and Other Histories of Sumatra.* Singapore: Singapore University Press, 2004.

———. *The Contest for North Sumatra: Atjeh, the Netherlands and Britain, 1858–1898.* Kuala Lumpur: Oxford University Press, 1969.

Rich, R., Edelstein, M., Hallman, M. and Wandersman, A. "Citizen Participation and Empowerment: The Case of Local Environmental Hazards". *American Journal of Community Psychology* 23, no. 5 (1995): 657– 76.

Smith, M. H. "American Religious Organizations in Disaster: A Study of Congregational Response to Disaster". *Mass Emergencies* 3 (1978): 133–42.

Snouck, H. *The Acehnese*. (trans. A. W. S. O'Sullivan), vols. I and II. Leiden: Brill, 1906.

Telford, J. and Cosgrave, J. "The International Humanitarian System and the 2004 Indian Ocean Earthquake and Tsunamis". *Disasters* 31, no. 1 (2007): S.1–28.

Tilley, C. *A Phenomenology of Landscape: Places, Paths, and Monuments*. Oxford: Berg, 1994.

Vale, L. and Campanella, T. *The Resilient City: How Modern Cities Recover From Disaster*. Oxford: Oxford University Press, 2005.

5

MANAGING POST-DISASTER RECONSTRUCTION FINANCE
International Experience in Public Financial Management

Wolfgang Fengler, Ahya Ihsan and
Kai Kaiser[1]

INTRODUCTION

The past decade has presented the development community with some of
its most demanding reconstruction challenges since the aftermath of World
War II. The World Bank and other development partners have been involved
in post-disaster reconstruction in response to the devastation resulting from
the tsunami in Indonesia (Aceh), Sri Lanka, the Maldives and India, and also
from the earthquakes in Pakistan and Indonesia (Yogyakarta/Central Java). The
World Bank and its partners have also supported post-conflict reconstruction
following peace agreements in Haiti and Sudan. All these activities came in
addition to other large-scale reconstruction programmes in Afghanistan, East
Timor and several other countries, most recently Lebanon.

In most cases, such disasters greatly exceed available domestic resources.
Consequently, international donor agencies are frequently called upon to
finance reconstruction in post-disaster and post-conflict countries. In the
case of large-scale natural disasters such as the Indian Ocean tsunami, private
contributions were also an important part of the reconstruction programme.

Spending these significant financial resources well has been a key concern in all these reconstruction episodes. Appropriate arrangements for Public Financial Management and Accountability (PFMA) are increasingly viewed as crucial ingredients to ensure that reconstruction proceeds with integrity in a timely and effective manner, while also adequately managing fiduciary risk.

The international community has increasingly emphasized the performance of Public Financial Management (PFM) systems to enhance the use of domestic resources in developing countries and to underpin the scaling up and effectiveness of aid. The strengthening of country financial management systems and donor harmonization have both emerged as key priorities in enhancing aid effectiveness, including through budget support. The recent Public Expenditure & Financial Management Accountability (PEFA) performance indicator framework has focused on benchmarking outcomes as a way of promoting capacity development in the PFMA area.[2]

This chapter focuses on special considerations for strengthening PFM arrangements in post-disaster and post-conflict reconstruction environments that have yet to receive systematic attention. This chapter's objective is twofold: (1) to present key features of PFM in post-disaster environments, and (2) to analyse the similarities and differences between PFM in post-disaster and post-conflict environments.

The application of sound fiduciary principles is very challenging in post-disaster situations, because the need for speed often overrides more conventional mechanisms for planning and implementing budgets. In addition, post-disaster and post-conflict situations often entail the engagement of many public and private development partners, necessitating that all parties work together effectively towards the objective of reconstruction. Mitigating the risk of corruption represents a crucial element in maintaining donor commitment and supporting the legitimacy of the overall reconstruction process.[3] We examine how recent PFM arrangements in six cases of post-natural disaster reconstruction — Indonesia (Aceh and Nias), Sri Lanka, Colombia, Grenada, Pakistan and the Maldives — have contributed to the management of reconstruction finance, highlighting key issues and considerations, together with a variety of approaches for strengthening these arrangements. Our approach seeks to adopt a more systematic assessment, such as comparing prioritization and sequencing of post-disaster PFM arrangements with conventional perspectives used in assessing PFM systems and processes. From a comparative perspective, this chapter also highlights similarities to and differences from purely post-conflict reconstruction, drawing selectively on examples in Afghanistan, East Timor, Haiti and Sudan.

The chapter first sets out the basic analytical framework for post-disaster/post-conflict responses and PFM cycles. Then it analyses recent country experiences against this framework. This chapter focuses on instances of post-natural disaster reconstruction, but also provides a comparative perspective on post-conflict situations. The final section presents some lessons on strengthening PFM arrangements for reconstruction based on the comparative experiences.

MANAGING RECONSTRUCTION FINANCE — AN ANALYTICAL FRAMEWORK

Post-disaster and post-conflict situations share a need to rapidly mobilize and deploy a significant level of public resources for relief and reconstruction. Whereas the relief phase is typically concerned with providing immediate support, the reconstruction phase typically involves a trajectory of returning to "normality". Recovery management includes the implementation of capital projects (for example, housing, schools and clinics), as well as re-establishing basic public services in a sustainable manner. The reconstruction phase is also subject to the prioritization of certain types of reconstruction, such as housing, livelihoods, and physical and social infrastructure.

Post-disaster and post-conflict situations also present the challenge of bridging the gap between the relief and reconstruction phases (Figure 5.1):

- Phase I is characterized by the relief effort and is typically led by the national government (often including the military), together with UN agencies. During this phase, which usually lasts several weeks, planning for reconstruction begins.
- Phase II presents the transition from an emergency to a full-scale reconstruction programme. Early reconstruction starts while emergency relief activities still continue. This is a critical phase for the success of the whole reconstruction programme. In many reconstruction programmes, the transition between emergency relief and reconstruction is poorly managed. This can create an unnecessary gap before reconstruction activities start and corresponding frustration among those affected. For example, frustration in post-tsunami Aceh ran high six months after the natural disaster, when core relief activities were phased out before most reconstruction activities had begun.
- Phase III represents the fully fledged reconstruction programme of which each component has its own sequence. For instance, in India the focus of

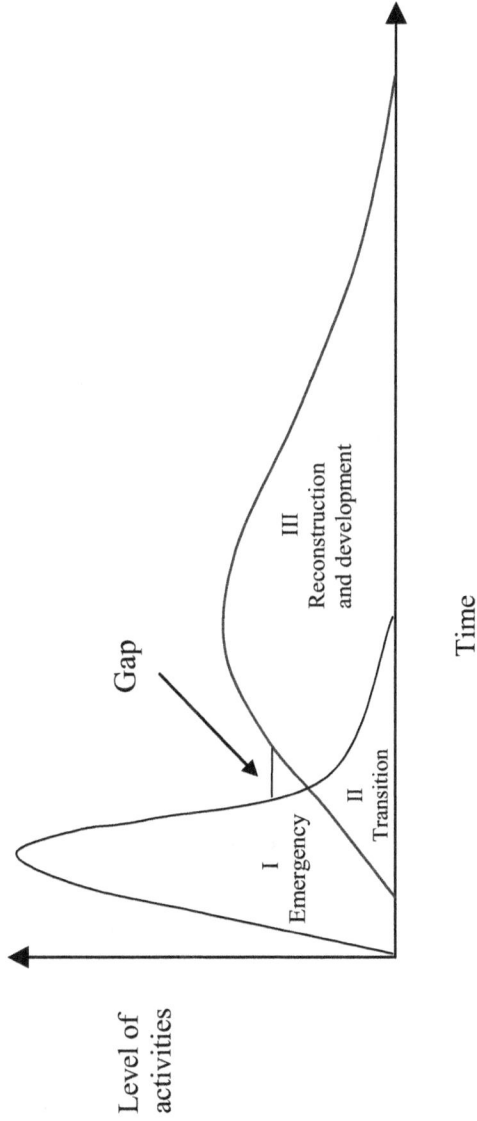

FIGURE 5.1
Implementation Phases of the Reconstruction Process

the first reconstruction year was on re-establishing livelihoods, particularly of affected fishing communities. By contrast, in Aceh and Nias, the first year was dominated by housing reconstruction, followed by a focus on infrastructure.[4]

POST-DISASTER AND POST-CONFLICT RECONSTRUCTION

While post-disaster and post-conflict reconstruction share the characteristics of immediacy and scale, notable differences exist. Natural disasters are typically unforeseen, while post-conflict reconstruction, often signalled by a peace agreement, offers some lead time. However, even in post-conflict situations, the call for reconstruction requires swift action, particularly when domestic and international resources aim to stabilize a fragile peace (Table 5.1).

Conflict often weakens the administrative and service-delivery capacity of states more than natural disasters. However, large-scale disasters, particularly if they affect a large proportion of a country, may also overwhelm in-country systems. Notably, post-conflict situations always carry the risk of unresolved political issues and a return to hostilities, making the reconstruction process fraught with uncertainty.[5] In the two most tsunami-affected regions, the coastal regions of Aceh and Sri Lanka, conflict and disaster overlap. However, in contrast to Sri Lanka, following the signing of the Helsinki Peace Accord on 15 August 2005, Aceh has been on track towards a durable peace settlement.

In post-conflict countries, the reconstruction challenge is often compounded by the need to rebuild a functioning public administration.

TABLE 5.1
Post-Natural Disasters vs. Post-Conflict Reconstruction

Similarities	Differences	
	Post-disaster	Post-conflict
• Donors need to respond fast, often with large volumes of aid • Set-up of new reconstruction agencies • Use of World Bank's financial instruments (for example, MDTF, ERL)	• Unforeseeable sudden event • Government system typically functioning regularly pre-disaster • More linear reconstruction path	• Often foreseeable • Government system associated with and often weakened by conflict • High likelihood of falling back into conflict

While the immediate priority is to re-establish key public services, another activity may be to strengthen the capacity of basic "core" public functions such as PFM and the civil service. These dual objectives are likely to create some trade-offs in terms of a focus on the strengthening and early use of government systems, against an early emphasis on other types of arrangement (including project implementation units, or PIUs).

In both post-disaster and post-conflict reconstruction, multi-donor trust funds (MDTFs) have emerged as one vehicle for channelling and coordinating reconstruction resources. MDTFs can be an effective and efficient tool in reducing transaction and management costs, particularly in high-risk environments. These typically entail a significant role for multilateral agencies such as the World Bank and the United Nations.

A recent review of eighteen MDTFs in post-crisis situations (post-disaster and post-conflict reconstruction) indicates that MDTFs have been an important instrument in resource mobilization, policy dialogue, and risk and information management. Since MDTFs frequently operate in high-risk and high-cost environments, they require flexible and adequate funding to enable effective and rapid responses to dynamic situations on the ground. MDTFs offer some advantages for national governments and donors in the post-crisis environment. They can increase and mobilize financial assistance and provide political visibility for the national authorities. Donors can operate in a more effective and efficient manner by reducing information, coordination and administration costs under joint financing arrangements. While MDTFs can be an effective tool in leveraging collective donor influence, they also entail less visibility for each individual donor.[6]

The immediate wake of a disaster or conflict typically comprises a number of stages that proceed in rapid succession: (1) damage/loss and needs assessments; (2) a donor conference; (3) the development of a reconstruction strategy (some elements of which may have already been presented at the donor conference); and (4) the coming into force of implementation modules and their integration into the budget cycle (Figure 5.2).

Damage/Loss and Needs Assessments[7]

Damage/loss and needs assessments have become vital instruments for governments and donors in estimating the level of damage, mobilizing resources and designing implementation arrangements. These assessments are often carried out by the host governments together with joint donor missions, typically led by the World Bank, the United Nations and/or regional development banks.

FIGURE 5.2
The Process of Mobilizing and Executing Reconstruction Finance

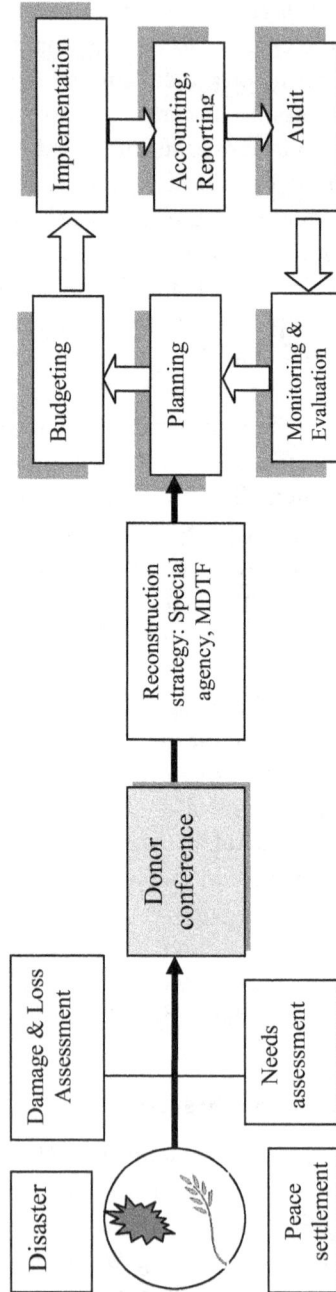

Damage/loss and needs assessments are related concepts, but their methodology is fundamentally different. Damage/loss assessments account for the loss of assets and the loss of flow of production of goods and services, as well as any temporary effects on the main macroeconomic variables subsequent to the event. Meanwhile, needs assessments calculate the financing requirements for reconstruction and do not necessarily reflect the damage/losses. Needs will greatly depend on the availability of resources, the duration of the recovery period and the government's own policies and priorities. Thus, the monetary figures for needs assessments can be higher or lower than those for damage/loss assessments, depending on the underlying determinants of need (Table 5.2).

TABLE 5.2
Needs Assessments versus Damage/Loss Assessments

Needs can be higher than damage/loss, because:	• Building back *better*, that is, beyond minimum services (that existed prior the disaster) • Emergency/transitional costs; incl. pure logistics of delivering reconstruction • Inflation
Needs can be lower than damage/loss, because:	• Fewer public services are needed in the case of large loss of life or migration • Insurance, the private sector or households cover part of the costs

Donor Conference

The impact of a disaster or conflict often far exceeds a developing country's capacity and resources to independently manage recovery. Financial assistance from international donors in these situations often plays a significant role. Donor conferences have become an important mechanism for mobilizing such international assistance. In such a forum, the preliminary estimates of damage/loss or needs assessment are presented, together with initial key policies of the government for directing the reconstruction.

Reconstruction Strategy

The preparation of a comprehensive reconstruction strategy includes decisions on the institutional and financial arrangements of the reconstruction programme. Depending on the scale of the disaster or the capacity of the national government, establishing a separate reconstruction agency is one option, particularly if the reconstruction effort receives continuous national and international attention. In this phase, governments typically also make decisions on establishing MDTFs.

PFM Strategy

The next important step is the implementation of the reconstruction programme in conjunction with the government's budget system. In most reconstruction episodes, particularly the larger ones, reconstruction financing flows both through the government's budget as well as outside the regular mechanisms ("off-budget"). In many cases, countries face a trade-off between rigorous planning and rapid action. The regular budget system is too rigid to allow for a sufficiently flexible response. In contrast, off-budget mechanisms face increased fiduciary risks and usually complicate coordination. The following section elaborates on the special considerations for reconstruction PFM.

Special Considerations for Reconstruction PFM

There is broad agreement that good PFM systems are essential for the implementation of policies and the achievement of development objectives, because good systems can support aggregate fiscal discipline, strategic allocation of resources and efficient service delivery. The Public Expenditure and Financial Accountability (PEFA) framework also provides benchmarks for PFM out-turns (budget credibility), cross-cutting features (comprehensiveness and transparency), the budget cycle (policy-based budgeting, predictability and control in budget execution, accounting, procurement, recording and reporting, and external scrutiny and audit), and donor practices.[8] The PEFA framework inspired some of the early actions, even in countries with weak capacity. These actions include the establishment of a consolidated budget (integrating both capital and recurrent expenditures), attempting to consolidate activities of multiple donors and ensuring that forward recurrent costs are covered.[9]

Good PFM practices apply in post-disaster and post-conflict reconstruction. For instance, the reconstruction budget should be credible, meaning that the resources promised for reconstruction are actually used for the intended purposes within a given time frame. However, post-disaster and post-conflict reconstruction efforts demand different treatment from regular budget cycles and procedures. This is because a speedier and more flexible response in a crowded environment of multiple actors is necessary. In short, reconstruction budgets often need to be drawn up from scratch, operate outside regular national or subnational budgets, and allow for speedier implementation of projects.

BOX 5.1
What Is Different about Reconstruction Budgeting?

- Speed. Reconstruction efforts are typically faced with significant time pressures. Progress is measured on a month-by-month basis, not an annual basis, as in regular projects. The need for a swift response means that the time periods for project preparation, budget approval and procurement need to be significantly shortened.
- Flexibility. Disasters or peace settlements rarely occur in sync with the budgetary process. In order to respond to such events, most governments have funds for immediate emergencies but often lack procedures needed to establish fast-track funding for immediate recovery. Once budgets are approved, emergency-recovery situations demand a greater flexibility to re-allocate funds within certain limits. In post-disaster environments, conditions change so rapidly that waiting until the national budget revision takes place would create unacceptable delays.
- Multiple actors. After large-scale disasters and high-profile conflicts, many government and non-government actors, often with limited expertise in the affected region or country, want to engage in the reconstruction. Aceh, a region that had been isolated before the tsunami, had more than 250 institutions supporting the reconstruction effort. These institutions often use different budget mechanisms to channel their

---

The main challenge in managing reconstruction finance is to integrate the specific reconstruction needs and conditions (speed, flexibility, multiple actors) into regular country systems in order to meet the highest fiduciary standards. In almost all reconstruction episodes, there is some degree of adjustment to the regular budgetary process (section III). The degree of this adjustment depends on the scale of the reconstruction effort, as well as the strength and flexibility of the respective country systems. Table 5.3 summarizes the main features of standard budget processes and their possible adjustment in reconstruction episodes.[10]

Reconstruction Strategy and Budget Cycle

Whereas securing sufficient resource commitments for reconstruction is an obvious priority in the wake of conflicts or natural disasters, the subsequent

TABLE 5.3
Regular and Reconstruction Budget Cycles

	Regular budget cycle (PEFA-principles)	Reconstruction financing
Planning	• Builds on past budget performance evaluation • Includes macro-framework based on economic outlook	• Builds on a damage/needs assessment • Needs to focus on rapid action while avoiding delays associated with standard annual planning cycles • Reconstruction agency/board may take on a special role
Budget Preparation	• Detailed vetting of projects by Ministry of Finance and line ministries • Approval by parliament	• Budgets are established from scratch • Needs high degree of flexibility, and anticipation of contributions from donors and NGOs • Standard unit costs often need to be revised upwards
Budget Execution	• Regular on-budget implementation	• Use of off-budget channels, particularly by UN and NGOs • Special procurement arrangements, including procurement agents
Accounting and Reporting	• Standardized and timely accounting and reporting of transactions	• Emphasis on comprehensiveness and transparency of the reconstruction budget, including off-budget flows
Audit and External Scrutiny	• A key fiduciary principle is that restriction funds are used for the purpose for which they were intended • Both *ex ante* and *ex post* controls will represent a critical element in ensuring that funds are not spent on unintended purposes	• Reliance on overly detailed *ex ante* controls will reduce flexibility and may risk delays in implementation • *Ex post* audit will be critical for assessing compliance and, in conjunction with adequate follow-up measures and sanctions, averting abuse
Monitoring and Evaluation	• Evaluates budget performance according to regular budget indicators. Examples include fiscal deficit, budget realism (implementation compared with original budget), and disbursement ratios	• Reliable information and analysis are even more critical in large reconstruction programmes than in regular development projects • Updates need to be more frequent, but real time tracking is unrealistic and not needed; quarterly updates of fund flows, reconstruction progress and basic economic indicators would be a major achievement

speed and integrity of reconstruction is the subject of increasing concern. Donors' commitments are themselves contingent on the assurance that resources will be well spent. International agencies, including the World Bank, have the credibility to assist in meeting these expectations and influencing other donors for fund mobilization. PFM is important because it creates a "credible environment" in which donors feel confident to make firm aid commitments, and it ensures that aid reaches the intended beneficiaries, helps governments strengthen fiduciary standards (including through demonstration effects) and garners stronger support from civil society organizations.

However, although core fiduciary principles apply, management, planning, budgeting and project implementation often need to follow different sequences and modalities in order to be effective in the early years of reconstruction. There are at least six decisions to make:

- Management and institutional set-up: Depending on the scale and location of the disaster, the size of the country and the strength of the local institutions, affected countries may set up independent reconstruction agencies (see also section 3.1). Alternatively, existing central or subnational government institutions typically coordinate the implementation of the reconstruction programme.
- Reconstruction planning versus rapid project implementation: A credible plan that includes key policy decisions is essential. However, lengthy planning exercises and overly detailed reconstruction plans can do more harm than good. Most importantly, the reconstruction process needs to start quickly, particularly to provide employment and livelihoods; plans can be readjusted along the way.
- On-budget versus off-budget: Fund flow arrangements highlight the tension between speed and orderly budget implementation in reconstruction programmes. International partners have increasingly emphasized the use of country PFM systems to channel aid, even in reconstruction situations. However, a large share of project implementation may be channelled outside regular budgetary processes, particularly if NGOs and the UN system are playing a significant role. This is not a major problem per se if a robust monitoring and evaluation system is in place. For funds that go through regular budgetary systems, it is critical to introduce a higher degree of flexibility and iterations of budgeting to allow for faster disbursements and reallocations. However, governments can achieve such flexible fund disbursements only if they have already established special fund-flow mechanisms that can also be used for the early reconstruction phase.

- Front-loading versus back-loading of funds: While quick action is essential, too much front-loading of reconstruction funds will likely increase inflation and reduce the resources available in the second and third years of reconstruction. The higher the share of NGO funding, the more funds the government can programme in later years, because NGO funding tends to be exhausted after two years.
- Regular versus special procurement regimes: Reconstruction procurement faces a dilemma. On the one hand, standard procurement processes need to be shortened to accelerate reconstruction. On the other hand, extra caution is needed because the influx of additional resources will put strain on the procurement system, which may be weak before disasters. A number of special or streamlined procurement arrangements have therefore been used, and a number of countries have also used independent procurement agents.
- Emphasis on *ex ante* or *ex post* controls: Reconstruction needs to strike a forward-looking balance between signalling a high degree of accountability while not allowing the implementation process to grind to a halt. The key decision is on the balance between *ex ante* and *ex post* controls. The more rapidly reconstruction needs to start, the more governments rely on *ex post* controls. However, the importance of these *ex post* controls then becomes even more significant than in regular development programmes.

Monitoring and Evaluation

Reliable information and analysis is even more critical in large reconstruction programmes than in regular development programmes. Multiple resource sources (from national governments, donors, NGOs) and implementation streams for reconstruction provide a particular challenge in tracking funds and evaluating results. In addition, the reconstruction PFM cycle is more compressed and iterative than regular budget cycles. The key benchmarks for the monitoring and evaluation (M&E) system are the production of timely and comprehensive estimates of (1) funds allocated and spent (covering all sources from domestic, public, international and private), (2) reconstruction progress, and (3) economic impacts.

Country Findings

The six natural-disaster country case studies reveal significant diversity in the scale of the reconstruction challenges and implementation modalities. The case studies include three post-tsunami countries (Aceh-Indonesia, Sri Lanka and Maldives); two post-earthquake countries (Colombia and Pakistan); and one post-hurricane country (Grenada). While all experienced severe damage/loss from natural disasters, Aceh and Sri Lanka had also suffered damage/loss from conflict. The extent of the national and international response, the type of intervention and the institutional and financial mechanisms all showed significant diversity among the six cases. This exercise relied on a mix of desk reviews and interviews with World Bank staff.[11]

For comparative purposes, this chapter also draws on purely post-conflict examples in Afghanistan, East Timor, South Sudan and Haiti. While East Timor's destruction was largely inflicted by human hand, it was of short duration. Such intensity can be compared with recent events in Lebanon, where destruction was concentrated in a short period akin to a natural disaster. In contrast, South Sudan's destruction as a result of civil war was far more protracted.

The exact nature and scope of the reconstruction challenge, combined with the country context, may contain important situational differences. Important considerations include: (1) the scale of international aid (public/private); (2) how much aid was channelled on-budget; (3) the country's institutional and budget arrangements; (4) how aid coordination was handled; (5) whether there were cross-cutting arrangements (for example, multi-donor trust funds); (6) how comprehensive and timely reporting was on commitments, fulfilled commitments and actual implementation/disbursement; (7) how many institutional levels were involved in the PFM cycle; (8) the financial arrangements for emergency relief and longer-term recovery; (9) what implementation arrangements were made; and (10) how fiduciary integrity/anti-corruption was managed for the various funding flows (including internal and external audit). These elements influence bottlenecks and the results achieved in the reconstruction process.

INSTITUTIONAL ARRANGEMENTS

All case studies involved the establishment of some type of special institutional arrangements — for example, a coordination body or special agency — to promote reconstruction. However, the extent to which these bodies engaged in coordination, monitoring and even implementation differed. All special agencies were given special roles in coordinating and monitoring, although other agents have often been left in charge of implementation.

With the exception of the Maldives, all case-study countries affected by natural disasters set up special reconstruction agencies. The post-conflict countries were also governed by special governance structures, either interim governments or a special power-sharing agreement as in the case of Sudan (Table 5.4). All the reconstruction agencies were assigned coordinating and monitoring functions, while the implementing tasks were mostly performed by the existing government line agencies or NGOs. Indonesia represents a special case where the reconstruction agency has authority to implement as well. However, in the first year, most of the implementation was undertaken by line ministries and local governments.

There are a number of compelling arguments for separating the functions of coordination and monitoring from implementation. First, since these agencies are typically created from scratch, they have no operational capacity or experience in implementation. Second, if experienced line agencies or local governments are bypassed, they may feel they have only a limited role in the reconstruction process. Third, establishing separate new bureaucracies for implementation may create incentives for these bureaucracies to be perpetuated and hinder their eventual phasing out. Finally, assigning the new agency with implementation could undermine its monitoring role and leadership.

There are several possible options with regard to the institutional arrangements for managing reconstruction (Table 5.5). Two commonly adopted models integrate a new agency into an existing ministerial system, usually in the form of a coordination body, or create a completely separate agency with specific authorities and responsibilities. The government may have strong leadership in post-disaster reconstruction and often plays a leading role. Conversely, in post-conflict settings the government role is frequently weak, requiring international intervention, for example, through the United Nations.

Colombia developed an innovative management model for managing reconstruction that has proven very successful. The reconstruction agency, FOREC, has a decentralized management structure with a clear distinction between national and local government functions. Its role is centred on coordinating and monitoring the overall recovery operation. Project implementation is carried out by NGOs at zonal management offices in thirty-two reconstruction zones through a competitive selection process. This model was successful in ensuring public participation, and social control and transparency, by contracting project monitoring to a consortium of universities while ensuring that the funds were administered by a fiduciary agency. The United Nations awarded FOREC the Sasakawa Prize on 11 October 2000 for its accomplishment in helping to prevent or reduce the risk from natural disasters.[12]

TABLE 5.4

Institutional Arrangements for Post-disaster and Post-conflict Reconstruction

Country	Type of event	Date of Event	Type of Institutional Arrangements	Implementing agency(ies)
Post-Disaster				
Indonesia (Aceh & Nias)	Tsunami and conflict	Tsunami: 26 December 2004; Peace Agreement: 15 August 2005	Special Agency: Badan Rehabilitasi dan Rekonstruksi (BRR)	Reconstruction agency, central government, provincial and local government, NGOs
Sri Lanka	Tsunami and conflict	Tsunami: 26 December 2004	Special Agency: Task Force for Rebuilding the Nation (TAFREN); Reconstruction and Development Agency (RADA)	Central government, provincial and local government, Donors, NGOs
Maldives	Tsunami	26 December 2004	Coordination Board: National Disaster Management Center (NDMC)	Central government, Donors, NGOs
Pakistan	Earthquake	5 October 2005	Special Agency: Earthquake Reconstruction and Rehabilitation Authority (ERRA)	Provincial and local government, partner organisation (PO), NGOs
Grenada	Hurricane Ivan	7 September 2004	Special Agency: Agency for Reconstruction and Development (ARD)	Central government, external partners and NGOs

			Special Agency	Reconstruction Agency, NGOs
Colombia	Earthquake	25 January 1999	Reconstruction Fund for the Coffee Region (FOREC)	Reconstruction Agency, NGOs
Post-Conflict				
East Timor	Conflict	30 August 1999 (Referendum)	Interim government 1999–2002: The United Nation Transitional Administration for East Timor (UNTAET)	Interim government, donors, NGOs
Afghanistan	Conflict	October 2001 (The fall of Taliban regime)	Interim government 2002–04 supported by The United Nation Assistance Mission in Afghanistan (UNAMA) and Coordination Agency: Afghanistan Assistance Coordination Authority (AACA)	Interim government, donors, NGO
South Sudan	Conflict	January 2005 (Peace Agreement)	Power-sharing arrangement between the North and the South. Presence of the UN (the United Nations Mission in the Sudan or UNMIS) Core Coordination Group (CCG) as coordination mechanism.	Government of Southern Sudan, NGO, private sector
Haiti	Conflict	March 2004, the fall of Aristide's regime	Interim Government 2004–06	Interim government

TABLE 5.5
Managing the Reconstruction Process: Institutional Options and Considerations

Institutional Options	Country	Advantages	Disadvantages	Consideration
Integration into existing ministerial system (Centralized Coordination Board)	Maldives Haiti	• Planning and budgeting, and oversight systems are in place • Established links with the international community • Sufficient capacity to implement	• The task may not be effectively addressed • Risk of lacking independence and local ownership/leadership • Civil service rules impede recruitment of professional staffs from outside	• Advisable in small countries or governments with established track record and strong administration to manage reconstruction
Separate reconstruction agency	Indonesia (Aceh) Sri Lanka Pakistan Grenada Colombia	• Most independent; and fully focused on reconstruction effort • Possible recruitment of new and dedicated team • The task can be effectively addressed	• Takes time to clarify roles and responsibilities of the agency • Possible disconnect from other government activities • May lack local ownership • Can take life of its own and is difficult to phase out	• Advisable in large scale and/or localized reconstruction episodes if agency has decision-making authority • Critical to establish sunset closure to avoid life of its own for too long

International intervention/support (interim government)	East Timor Afghanistan	• Ensuring the existence of government and administration • Access to international community	• May lack local support • Lack of understanding of local needs	• An option if the government and administrative functions have collapsed • This option is preferred only for a short-term transition
Integration into existing provincial and local government structures (Decentralized Coordination Board)	India (Tsunami) Indonesia (Yogyakarta, Central Java)[1]	• Full local ownership • Rehabilitation of sectors corresponds to most of the decentralized functions if country is very decentralized	• Provincial and local governments could be overburdened due to (1) losses from the disaster, and (2) inadequate capacity to manage a large reconstruction programme • Civil service rules impede recruitment from outside • Potential fiduciary risks, as existing government system may not adequately address reconstruction challenges	• Advisable if reconstruction effort is of manageable scale, local governments are strong and already empowered through decentralization • If central agencies lead the reconstruction process in the early phase, local governments should gradually play an increasing role

FOREC faced a number of institutional challenges. The exclusion of local governments from project identification and implementation may have resulted in their lack of full cooperation. Capacity and engineering skills were dissimilar across the thirty-two zone managers, resulting in implementation divergences in some areas. This decentralized function also created challenges with coordination and information collection.

The other management model is Pakistan's Earthquake Reconstruction and Rehabilitation Agency (ERRA). ERRA was established as an independent agency to coordinate, monitor and support the government in the implementation of reconstruction projects. It has decentralized offices at the province and district levels. ERRA is staffed by a professional, dedicated team and provides technical assistance to provincial and district governments as an implementing agent for budget preparation, procurement and project implementation.

In exceptional cases, when official government structures are weak or non-existent, the UN has acted as an interim government. In East Timor, the UN administered the transitional government and managed the initial phase of reconstruction, while at the same time ensured that the government system continued to function. The United Nations Transitional Administration for East Timor (UNTAET) acted as the coordinating and monitoring agent. The implementation function was largely carried out by international donors and international NGOs, with a limited role for local governments.[13]

FINANCING AND EXECUTION ARRANGEMENTS

A central feature of a reconstruction experience is its reliance on multiple sources of support. Government, multilateral, bilateral and non-governmental agencies all contribute to the reconstruction process. For example, in Aceh, Sri Lanka, East Timor and Afghanistan, international donors and non-governmental actors played a significant role in the initial relief/rehabilitation process.

The mix of public and private funds differs from case to case. The tsunami triggered one of the largest mobilizations of private funds in development history. People and governments around the world participated in an unprecedented act of global solidarity. Private contributions reached record highs, estimated at more than US$10 billion for emergency support and reconstruction programmes. In Aceh and Sri Lanka, the NGO sector became one of the main contributors to the reconstruction efforts, and its funds have financed most of the existing reconstruction activities so far. In Aceh alone, NGOs implemented programmes worth almost US$2 billion (as of end April 2006) and are expected to finance about thirty per cent of the total reconstruction programme.

NGOs (and, within limits, the UN) can react faster to deliver emergency supplies and reconstruction in their sectors of comparative advantage (often social sectors). Classical (and larger) donors and national governments (depending on their levels of income) are slower but have a comparative advantage in bulky, large-scale and complex investments, particularly in infrastructure.

Due to their specialization and comparative advantage, NGOs and many donors can pre-programme their funds very rapidly. This early planning and programming helps implementation. However, the complexities of reconstruction and the multiplicity of players can create special challenges for project coordination and management, especially in situations where off-budget arrangements are prominent. It is critical that sufficient fungible funds remain available for reconstruction gap-filling and development programmes in the second reconstruction phase.

Joint financing arrangements such as MDTFs have been used by most countries to improve the coordination and effectiveness of reconstruction processes. This model is important in situations where the bulk of resources come from bilateral and multilateral donors. The World Bank has been both a trustee and administrator of MDTFs. In addition to enhancing effectiveness of coordination, this arrangement increases donor confidence by providing assistance where the fiduciary systems of a country are weak.

There are a limited number of cases where the government manages such joint financing arrangements. This requires sound government budget systems and a number of fiduciary risk-measurement indicators to be in place to ensure accountability and transparency and establish donor confidence. The Maldives established an MDTF administered by the government, while in Pakistan the government dedicated a single-basket account as a reconstruction fund, which is managed by its reconstruction agency (Table 5.6).

Post-conflict reconstruction poses a special challenge in financial management given the insecure environment. Simultaneously weak administrative procedures and national budget systems receiving large external financial assistance may result in the diversion of investments away from national development priorities. Often, governments do not have adequate capacity to coordinate and monitor pledges, expenditures or outcomes. In such settings, joint donor financing instruments such as MDTFs can play a vital role in consolidating external finances and improving coordination and effectiveness in implementation.

TABLE 5.6
Natural Disaster and Post-conflict World Bank Financing Modalities

Country	MDTF	New ERLs	Reprogramming of portfolio
Post-Disaster			
Aceh and Nias	√		√
Sri Lanka		√	√
Maldives	√	√	√
	(government-administered)		
Pakistan		√	√
Grenada	√	√	
Colombia	√		
Post-Conflict			
East Timor	√		
Afghanistan	√		
South Sudan	√		
Haiti		√	

The complexity of aid financing coordination can create barriers to national ownership of the reconstruction planning process and prevent the integration of all funding sources into the national budget. In East Timor, donor financing arrangements were complex and channelled through four different mechanisms: the UN humanitarian consolidated appeal, the UNTAET budget, the CFET-UN administered trust fund, and the TFET-World Bank-administered trust fund.[14]

Implementation Experience

Successful reconstruction processes include responsiveness to needs/effective prioritization, timeliness, cost effectiveness (given time constraints) and the assurance that funds will be spent on their intended purpose in a timely manner (including the suitable management of corruption risks). Key benchmarks that PFM systems can influence in a significant way are timely, credible information; timely and equitable implementation; and efforts to minimize corruption (Table 5.7). The design and implementation of PFM systems contribute to achieving these results but they are not the single determinant. Many other factors also come into play, particularly the availability of resources, the availability and price of construction materials, the capacity to manage and implement reconstruction programmes, and the level of coordination between all reconstruction parties.

TABLE 5.7
Selected Performance Criteria for Reconstruction PFM

Criteria	Results measures	Key considerations	Issues
Information	• Timely and comprehensive information on commitments and disbursements of reconstruction resources • Comprehensive financial information • Physical progress	Critical in reconstruction contexts given the need for timely information on: (1) Financial flows (government, donors, NGOs) (2) Physical progress (3) Economic impact	Information piecemeal, tendency to highlight easy information, including own agency (part incentive for self-advertisement)
Implementation (effectiveness and efficiency)	Timely implementation of reconstruction according to prioritization: (1) Timely allocations: planning (2) Timely commitments: procurement and contracting (3) Timely implementation: disbursements	Case studies should reveal how different processes worked. Did "needs" match commitments in different areas?	Did countries seem to get initial prioritization process right? Was this translated into execution? When setting up reconstruction agencies (for example, Aceh, Pakistan) there is a trade-off between slowing down reconstructions in the early phase in exchange for a more cohesive reconstruction programme in the medium term
Implementation (equity)	Equity in implementation of reconstruction	Rural/low-capacity areas may be relatively neglected in reconstruction process	Evidence that reconstruction was bunched in capital
Anti-corruption	Effective management of fiduciary risk/corruption minimized	• Where effective safeguards in place? • Did these vary significantly across funding flows? • Was an effective balance struck between speed of implementation and management of fiduciary risk?	In Aceh, the reconstruction agency cancelled a number of projects considered to have unfair bidding processes. There is evidence that some government officials are reluctant to be project managers due to tight anti-corruption procedures. This slowed implementation.

Information

Financing is likely to come from multiple sources. Timely and consistent tracking of budgets/commitments and execution is a critical ingredient for assessing reconstruction progress. Financial information is especially sensitive, particularly if donors and private contributions are high. An ongoing assessment of commitments and disbursements across all sources (public/private, domestic/international) would highlight the extent of any reconstruction delays and potential financing gaps. Good PFM systems would then measure whether the financial resources are being translated into outcomes. When execution is decentralized across various types of domestic and international agencies or local governments, data systems need to pay particular attention to proper accounting. For instance, commitments need to be separated from disbursements, emergency projects from reconstruction projects, and financing institutions from implementing partners.[15]

In recent years, several "tracking systems" have been developed to monitor financial information and improve aid management of the recovery process. The existence of credible and integrated financial tracking systems has become more critical in recent years, because of the unprecedented levels of funding after the tsunami, together with other large-scale post-conflict reconstruction efforts (for example. Afghanistan and Sudan) with large funding.

The most prominent aid management and reconstruction tracking instrument is the Development Assistance Database (DAD). The DAD has been applied in a number of countries including Afghanistan, Sri Lanka, the Maldives and Indonesia. However, coverage has been limited, often focusing on UN and other core donor activities. Although the DAD includes sophisticated technical specifications, such as visual breakdowns to the village level, it has faced many implementation challenges, particularly in Aceh and Nias.

While information technology has sometimes been an impediment to the effective implementation of information systems, it has typically not been the main obstacle for effective monitoring. Instead, the main challenges are related to data collection and analysis:

- In Aceh and Nias, a labour-intensive data collection effort by a joint team of the World Bank (mainly national staff) and the reconstruction agency has proven superior to DAD's more high-tech and self-entry-based information system in tracking funds. The Indonesia DAD, a more than US$2 million investment, has yet to deliver any significant results.
- In Sri Lanka, DAD provides regular financial and project updates. Similar to Aceh and Nias, this system also builds on self-entry-based

information. In order to increase compliance, the DAD team also discloses the institutions not providing data.

Future technical assistance should pay particular attention to more basic but effective systems. For example, the DAD system relies mainly on self-entry. A better but admittedly more labour-intensive approach is to engage in proactive collection from key players. Furthermore, rather than focusing on excessive detail, a more straightforward classification system that focuses on the core sectors of the damage/loss assessment would generate more workable systems.

Implementation

Experience reveals a trade-off between swiftness and national ownership (capacity-building) in reconstruction implementation. Physical reconstruction, to some extent, is less problematic and faster to implement than institutional capacity-building. Sequencing is critical in the situation of both low institutional capacity and widespread needs. In East Timor, sectors that made progress in establishing institutions and strong levels of management capacity such as health were often slower initially in achieving physical reconstruction targets.[16]

In weak and/or cumbersome governance arrangements, off-budget channels (partially through NGOs) seem to be critical in the early phase of reconstruction. In Aceh and East Timor, mobilization of the private sector and NGOs at the initial stages, combined with community development-driven reconstruction, achieved rapid results on the ground and also increased community participation (for example, cash-for-work programmes).

In post-conflict situations, security remains a major challenge in the recovery process. In Afghanistan, implementation of the development programme has been constrained by poor security conditions. In many cases, despite the rapid approval of projects and budget allocations from donors and government, implementation itself has been slow. In addition, limited capacity in line ministries, in the areas of procurement and financial management, has been the major problem faced by Afghanistan. In East Timor, the implementation of large reconstruction projects was slow, largely due to difficulties in managing standard procurement procedures in a post-conflict context.

There has been increasing concern over the equitable geographic distribution of funds. The most accessible areas, usually the regional capital, invariably receive more funding than they need. Conversely, isolated areas often lag behind in funding. In Aceh, there has been a bias towards the areas closer to the capital city, Banda Aceh, which received double its needs. In Afghanistan, poor security prevents donors and NGOs from providing assistance to isolated areas, thus centring reconstruction in Kabul.

LESSONS LEARNED

This chapter has sought to offer an initial stock-take of PFM experience in reconstruction. All those cases examined highlight three main lessons that are key to successful reconstruction: (1) the fundamental difference between reconstruction programmes and regular development programmes, particularly with regard to the swiftness of action, (2) the importance of adjusting the standard budget cycle while protecting core fiduciary principles, and (3) good information and communication.

Reconstruction programmes are different from regular development programmes: Swift action is just as important as careful planning.
In the early phase, swift action is the overriding concern and a rapid response can only be guaranteed with a high degree of flexibility in PFM arrangements. Many governments have provisions for emergency response but lack budget flexibility in the early reconstruction phase (three to twelve months into the reconstruction programme). Once a significant portion of the reconstruction programme has been delivered, it is crucial to manage the transition back to regular budget processes. In many cases, local governments are key but often underestimated players in sustaining the capital investments made during the reconstruction process.

It is important to identify the changing environment of PFM arrangements early on, in order to allow the home government and donors to adjust accordingly. In cases where a new reconstruction agency has been set up, the risk of slow project implementation can be mitigated by allowing for off-budget execution or substantial back-loading of the government's own contributions. Many players, particularly NGOs, bilateral donors and the UN, dedicate their funds to specific sectors and focus on the early reconstruction phase. Fungible resources for reconstruction and development are very important in the second reconstruction phase. The affected governments and/or larger multilateral donors have a comparative advantage in providing these fungible programmes.

The core principles of good PFM apply in both post-disaster and post-conflict reconstruction, but not the standard annual budget cycle.
To avoid delays in implementation, reconstruction needs to start immediately. Hence, associated PFM sequencing needs to identify core emergency procedures and minimal functionality to avoid implementation delays. Rather than waiting to put in place a fully fledged system and comprehensive planning strategy, critical projects may require emergency modalities.

While on-budget arrangements are preferred for long-term development, off-budget mechanisms have been effective in responding to emergency needs and allowing for more flexibility in rapidly changing circumstances. Coordination and integrated monitoring of projects should avoid any overlap and waste resources.

Good information and communication are the secret of successful reconstruction: The combination of large amounts of funding and the need for rapid action creates an environment where reliable evaluation and monitoring is critical to success.
Many important decisions, particularly funding decisions, are taken at short notice and based on weak information. An integrated accounting and reporting mechanism that covers both off- and on-government budget contributions is critical, if the use of reconstruction resources is to be better planned and managed. In many cases, off-budget funds, particularly from NGOs, are not part of the overall accounting and reporting system. Special arrangements, such as project approval workshops as in the case of Aceh and Nias, are needed to capture these off-budget flows.

To evaluate overall progress and outcomes, regular monitoring should be conducted jointly by governments, international agencies and NGOs. Regular and systematic updates of key progress indicators are critical for informing and improving subsequent (funding) decisions and adapting the reconstruction programme. In many cases, monitoring and evaluation are not conducted comprehensively. Instead, most M&E systems focus only on the coordinating agency's own performance rather than the recovery and reconstruction performance as a whole.

Despite ambitious pronouncements, existing high-tech systems such as the UN's DAD have as yet not been able to provide timely and credible information that can be used for policy decisions. The experience of the reconstruction effort in Aceh and Nias shows that relatively low-tech, labour-intensive data collection and analysis based on a robust methodology is superior to more sophisticated and self-entry based information systems.

Notes

1. An updated version of this chapter has been published as the World Bank's Policy Research Working Paper WPS4475. An early draft of this chapter was presented to The First International Conference on Aceh and Indian Ocean Studies, February 26, 2007, Banda Aceh. We are grateful for the useful comments made by Anand Rajaram, Saroj Kumar Jha, Zoe Trohanis, Francis Ghesquiere, Amitabha Mukherjee, James Brumby, Robert Floyd, and all country team colleagues who provided information and insights on the country cases. Cut Dian 'Agustina, Peter Milne, and Stefan Nachuk have also provided significant inputs into this paper.
2. The multi-donor PEFA initiative performance measurement framework covers twenty-eight indicators, with an additional three to assess donor practices (see The PEFA Public Financial Management Performance Measurement Framework, June 2005) across the full budget cycle.
3. Recent PFM work has also focused on identifying appropriate reform frameworks for low capacity contexts. For example, the platform approach to PFMA is being piloted in a number of countries, as reforms to date have often proved unwieldy, poorly coordinated and unsustainable. The approach aims to achieve increasing levels ("platforms") of PFMA competence over a manageable time frame. Each platform establishes a clear basis for launching the next, based on the premise that a certain level of PFMA competence is required to enable further progress to take place; see Department for International Development (DFID), *A Platform Approach to Improving Public Financial Management*, Policy Discussion Briefing (July 2005) <http://www.cipfa.org.uk/international//download/Briefing_Platform_July05.pdf>.
4. Badan Rehabilitasi dan Rekonstruksi (BRR), *One Year after the Tsunami: The Recovery Effort and Way Forward* (Jakarta, Indonesia: NAD-Nias and Internasional Partners, 2005).
5. Estimates are that about half of peace agreements are associated with a reversion to conflict; see Paul Collier, Anke Hoeffler and Mans Soderbom, *Post-Conflict Risks,* CSAE WPS/2006-12 (Oxford: Center for the Study of African Economies, Department of Economics, University of Oxford, 2006).
6. *Review of Post-Crisis Multi-Donor Trust Funds* (Norway: Scanteam, February 2007).
7. Damage/loss assessments are mostly applied in the aftermath of natural disasters while needs assessments are frequently used to estimate financing requirements in the post-conflict situation. The Economic Commission for Latin America and the Caribbean (ECLAC) has developed a standard methodology to assess damage and losses after natural disasters; see *Handbook for Estimating the Socio-Economic and Environmental Effects of Disasters* (New York/Washington DC: Economic Commission for Latin America and the Caribbean, 2003). Needs assessments take a broader costing approach and include institutional, policy and infrastructure

needs. In post-conflict reconstruction, needs assessments predominantly focus on the costs of state-building; see *Practical Guide to Multilateral Needs Assessments in Post-Conflict Situations* (New York: UNDP/UNDG/World Bank, 2004).

8. The PEFA Public Financial Management Performance Measurement Framework (June 2005).

9. Feridoun Sarraf, *Integration of Recurrent and Capital "Development" Budgets: Issues, Problems, Country Experiences, and the Way Forward.* Paper prepared for Public Expenditure and Financial Accountability (PEFA) Program (Washington DC: World Bank, 2005).

10. See Annex 1 for a full adaptation of the PFM framework to reconstruction programmes.

11. See Annex 2 for a summary of the reconstruction profile of the ten country cases.

12. World Bank, *Implementation Completion Report on a Loan in the Amount of US$225 million to The Government of Colombia for the Earthquake Recovery Project* (Washington DC, 2003).

13. For example, in Afghanistan and East Timor, a lack of capacity has meant that day-to-day budget activities are carried out by international consultants; see William Dorotinsky and Shilpa Pradhan, "Exploring Corruption in Public Financial Management", in *The Many Faces of Corruption: Tracking the Vulnerabilities at the Sector Level*, edited by J. Edgardo Campos and Sanjay Pradhan (Washington DC: World Bank, 2006).

14. Sarah Cliffe and Klaus Rohland, *The East Timor Reconstruction Program: Successes, Problems, Tradeoffs*, Working Paper (Washington, DC: World Bank, 2002).

15. Cut Dian R. D. Agustina, *Experience with Reconstruction Finance Tracking Systems* (Jakarta: Technical Note, 2007).

16. Cliffe and Rohland. *East Timor Reconstruction Program.*

ANNEX 5.1
Framework: Conventional and Post-Disaster/Post-Conflict PFM Cycle

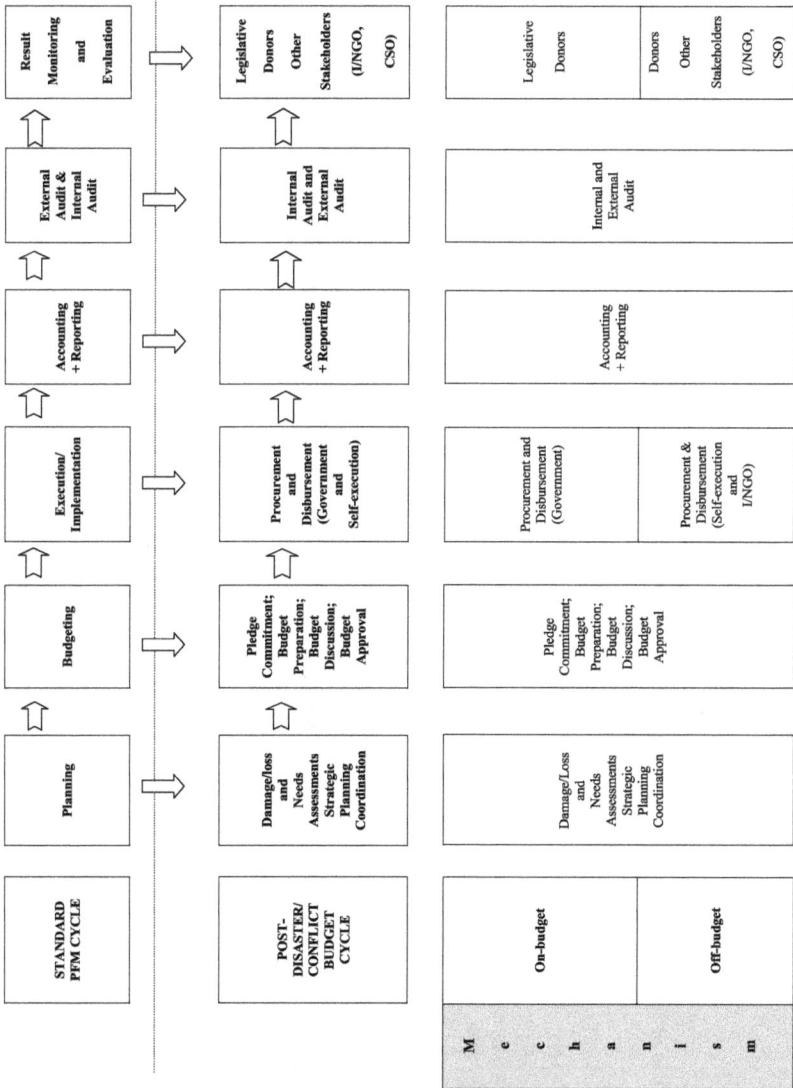

Mechanism	STANDARD PFM CYCLE	Planning	Budgeting	Execution/ Implementation	Accounting + Reporting	External Audit & Internal Audit	Result Monitoring and Evaluation
	POST-DISASTER/ CONFLICT BUDGET CYCLE	Damage/loss and Needs Assessments Strategic Planning Coordination	Pledge Commitment; Budget Preparation; Budget Discussion; Budget Approval	Procurement and Disbursement (Government and Self-execution)	Accounting + Reporting	Internal Audit and External Audit	Legislative Donors Other Stakeholders (I/NGO, CSO)
On-budget		Damage/Loss and Needs Assessments Strategic Planning Coordination	Pledge Commitment; Budget Preparation; Budget Discussion; Budget Approval	Procurement and Disbursement (Government)	Accounting + Reporting	Internal and External Audit	Legislative Donors
Off-budget				Procurement & Disbursement (Self-execution and I/NGO)			Donors Other Stakeholders (I/NGO, CSO)

ANNEX 5.2
Country Cases: Key Facts

Country	Date of Event	The Nature and Impact of Reconstruction Challenge	Date of Damage and Loss or Needs Assessment	Donor Conference
Indonesia (Aceh and Nias)	Indian Ocean Tsunami: 26 Dec. 2004 The signing of peace agreement: 15 August 2005 (after 30 years of conflict)	The impact of tsunami: • 130,000 people killed • Damaged and losses: US$4.5 billion • Impact on economy: 97% of province GDP or 2% of national GDP 15,000 killed by conflict	18 Jan. 2005	Jakarta, 19 Jan. 2005 Donor pledges: US$5.1 billion (grants & loans)
Sri Lanka	Indian Ocean Tsunami: 26 Dec. 2004 Ceasefire in 2002, the conflict is still ongoing	The impact of tsunami: • 35,322 people killed • Damage and losses: US$2.2 billion • Impact on economy: 7.6% of GDP More than 20 years of conflict: displaced 390,000 people	10 Jan. 2005. The report was released on 2 Feb. 2005	Jakarta, 19 Jan. 2005 Donor pledges: US$2.1 billion (grants & loans)
Maldives	Indian Ocean Tsunami: 26 Dec. 2004	• 29,000 people displaced • Damage and losses: US$470 million • Impact on economy: 62% of GDP	Early Jan 2005. The report was released on 8 Feb. 2005	Jakarta, 19 Jan. 2005 Donor pledges: US$302 billion (grants & loans)
Pakistan	7.8 Richter scale earthquake on 5 Oct. 2005	• 73,000 people dead • Damage and losses: US$5.2 billion • Impact on economy: 0.4% of GDP (exc. Azad Jammu and Kashmir/AJK)	12 Nov. 2005	Tokyo, 19 Nov. 2005. Donor pledges: US$5.9 billion (grant + loan)

continued on next page

ANNEX 5.2 — cont'd

Country	Date of Event	The Nature and Impact of Reconstruction Challenge	Date of Damage and Loss or Needs Assessment	Donor Conference
Grenada	Hurricane Ivan: 7 Sept. 2004	• Damaged 80% of public and private infrastructure • Damage and losses: US$800 million • Impact on economy: 200% of GDP	13 Sept. 2004	Washington, DC, 4 Oct. 2004 Donor Pledges: US$150 million
Colombia	6.2 Richter scale earthquake on 25 Jan. 1999	• 1,185 people killed • Damage and losses: US$1.86 billion • Impact on economy: 2.2% of GDP	April 1999	None
East Timor	Mass violence and clash between pro-Indonesia and pro-independence groups after referendum in 30 Aug. 1999 resulted in infrastructure destruction and the collapse of the state structure	• Displaced 75% of population and destroyed 70% of infrastructure • Need assessment: US$307 million • Impact on economy: 40–45% drop in GDP	Oct.–Nov. 1999	Tokyo, Dec. 1999 Donor pledges: US$366 million
Afghanistan	The fall of the Taliban regime on October 2001	• Millions of people displaced • Needs assessment: US$14.6 billion (10 years) • GDP 2002: US$4.4 billion	Jan. 2002	Tokyo, 21–22 Jan. 2002 Donor pledges: US$5.1 billion
South Sudan	Signing of Peace Agreement on January 2005 after 20 years of civil war	• 2 million people killed • Needs: US$3.6 billion (or US$8 billion in total)	Dec. 2003–18 March 2005 (Joint Assessment Mission)	Oslo, April 2005 Donor pledges: S$4.5 billion
Haiti	March 2004, the fall of Aristide's regime, after over two decades conflict	• 65% of population below poverty line • Needs assessment: US$1.4 billion • Impact on economy: 5.5% of GDP	July 2004 (Interim Cooperation Framework 2004–2006)	Washington DC, July 2004. Donor pledges: US$1 billion

Notes:
1. The experiences of India's post-tsunami reconstruction and Indonesia's post-earthquake reconstruction of Yogyakarta and Central Java (27 May earthquake) are not part of the ten case studies. However, in both cases, the national governments decided to empower the subnational governments

ANNEX 5.3
Country-cases

Indonesia (Aceh and Nias)
Badan Rehabilitasi dan Rekonstruksi (BRR) NAD-Nias and International Partners. 2005. *One Year after the Tsunami: The Recovery Effort and Way Forward.* Jakarta, Indonesia

Fengler, Wolfgang and Ihsan, Ahya 2006. Tracking the Money. 10 Lessons from the Aceh and Nias Reconstruction Effort. Presentation at PREM week. May 2006, Washington D.C.

World Bank and Badan Rehabilitasi dan Rekonstruksi (BRR) NAD-Nias. 2005. *Rebuilding a Better Aceh and Nias: Stocktaking of the Reconstruction Effort.* Jakarta, Indonesia

World Bank. 2007. *One Year after the Java earthquake and Tsunami: Reconstruction Achievements and the Results of the Java Reconstruction Fund.* Jakarta, Indonesia

Sri Lanka
Government of Sri Lanka. 2005. *Post Tsunami Recovery and Reconstruction: Progress, Challenges, Way Forward.* Colombo, Sri Lanka.

Government of Sri Lanka and Development Partners. 2005. *Post Tsunami Recovery and Reconstruction: Progress, challenges, way forward.* Colombo, Sri Lanka.

Reconstruction and Development Agency (RADA). 2005a. *Establishment of the Reconstruction and Development Agency (RADA)* available at: <http://www.tafren.gov.lk>.

Reconstruction and Development Agency (RADA). 2005b. *The Role of TAFREN in Post-Tsunami Recovery in Sri Lanka* available at: <http://www.tafren.gov.lk>.

World Bank. 2005b. Technical Annex for a Proposed IDA Grant in the Amount of SDR 20.1 million (US$30 million equivalent) and a Proposed IDA Credit in the Amount of SDR 30.2 million (US$45 million equivalent) to the Democratic Socialist Republic of Sri Lanka for a Tsunami Emergency Recovery Programme — Phase II. Washington D.C.

Maldives
Government of Maldives. 2005. *The Maldives: One Year after the Tsunami.* Male, Maldives.

Government of Maldives, Ministry of Finance and Treasury. 2005. *Tsunami Relief and Reconstruction Fund: Operation Manual.* Maldives

World Bank. 2005c. Technical Annex for a Proposed IDA Grant in the Amount of SDR 3.7 million (US$5.6 million Equivalent) and a Proposed IDA Credit in the Amount of SDR 5.6 million (US$8.4 million Equivalent) to the Republic of Maldives for a Post-Tsunami Recovery and Reconstruction Project. Washington D.C.

The official website of National Disaster Management Center (NDMC). 2005. Available at <http://www.tsunamimaldives.mv>.

Pakistan
Asian Development Bank and World Bank. 2005. Pakistan 2005 Earthquake: Preliminary Damage and Needs Assessment. Islamabad, Pakistan. World Bank. 2005d. Technical Annex for a Proposed Credit in the Amount of SDR 281.8 Million (US$400 Million Equivalent) to the Islamic Republic of Pakistan for an Earthquake Emergency Recovery Credit. Washington D.C.

United Nations. 2005. *Pakistan 2005 Earthquake: Early Recovery Framework*. Islamabad, Pakistan.

World Bank and Asian Development Bank. 2005. *Pakistan 2005 Earthquake: Preliminary Damage and Needs Assessment.* Islamabad, Pakistan.

Grenada
Organization of Eastern Caribbean States. 2004. *Grenada: Macro Socio-Economic Assessment Caused by Hurricane Ivan.* www.oecs.org

World Bank. 2004a. *Grenada Hurricane Ivan: Preliminary Assessment of Damages.* Washington D.C.

World Bank. 2005e. *Grenada: A Nation Rebuilding: An Assessment of Reconstruction and Economic Recovery One Year after Hurricane Ivan.* Washington D.C.

Honduras
World Bank. 2004b. *Learning Lessons from Disaster Recovery: The Case of Honduras.* Washington D.C.

Colombia
World Bank. 2003. Implementation Completion Report on a Loan in the Amount of US$225 million to The Government of Colombia for he Earthquake Recovery Project. Washington, D.C.

"ECLAC Study on Impact of January Earthquake in Colombia to be Presented to President Andres Pastrana". Available at <http://www.scienceblog.com/community/older/archives/L/1999/A/un990625.html>.

East Timor
Cliffe, Sarah and Rohland, Klaus. 2002. The East Timor Reconstruction Program: Successes, Problems, Tradeoffs. Working Paper. World Bank, Washington D.C.

Alonso, Alvaro and Brugha, Ruairi. 2006. *Rehabilitating the Health System after Conflict in East Timor: A Shift from NGO to Government Leadership.* Oxford University Press, London.

Afghanistan
Scanteam. 2005. *Assessment, Afghanistan Reconstruction Trust Fund.* Oslo, Norway.

Suhrke Astri et al. 2002. *Peacebuilding: Lessons for Afghanistan.* Chr. Michelsen Institute. Norway.

World Bank. 2005f. *Afghanistan: Managing Public Finances for Development. Main Report (Vol. 1).* Poverty Reduction and Economic Management Sector Unit South Asia Region, Washington, D.C.

Mckechnie, Alastair. 2003. Humanitarian Assistance, Reconstruction and Development in Afghanistan: A Practitioner's View. CPR Working Paper. Washington, D.C.

Asian Development Bank, United Nations Development Programme, and World Bank (ADB, UNDP, and WB). 2002. *Afghanistan: Preliminary Needs Assessment for Recovery and Reconstruction.*

South Sudan
Joint Assessment Mission (JAM) Sudan. 2005a. *Framework for Sustained Peace, Development and Poverty Eradication. Volume 1: Synthesis.* Available at <http://www.unsudanig.org/JAM/>.

Joint Assessment Mission (JAM) Sudan. 2005b. *Framework for Sustained Peace, Development and Poverty Eradication. Volume 2: Cluster Costings and Matrices.* Available at <http://www.unsudanig.org/JAM/>.

Joint Assessment Mission (JAM) Sudan. 2005c. *Framework for Sustained Peace, Development and Poverty Eradication. Volume 3: Cluster Reports.* Available at <http://www.unsudanig.org/JAM/>.

Haiti
World Bank. 2004c. Haiti Country Overview. Available at <http://www.worldbank.org/ht>.

Republic of Haiti. 2004. Summary Report — Interim Cooperation Framework 2004–2006 Haiti.

Taft-Morales, Maureen. 2005. *Haiti: International Assistance Strategy for the Interim Government and Congressional Concerns.* Congressional Research Service.

6

BETWEEN CUSTOM AND LAW
Protecting the Property Rights of Women after the Tsunami Disaster in Aceh

Daniel Fitzpatrick

INTRODUCTION

Few would dispute the potential for natural disasters and armed conflicts to have a disproportionate impact on women, especially in the developing world. Women who are primary caregivers, with greater responsibility for household work, have less time and capacity to mobilize resources for recovery. They are less likely to participate in the public sphere in which humanitarian relief is organized and delivered. They may be overlooked if relief efforts target programmes at household heads, or focus on primary employment as the sole source of livelihoods. And if these relief efforts also fail to collect gender-disaggregated data, the disproportionate impacts on women may not even register in monitoring mechanisms.[1]

Population displacement induced by disaster or conflict can remove women from kinship structures that provide basic forms of social insurance against poverty and violence. Displacement also removes women from location-specific income, including access to common property resources. After displacement, women who return home are at risk from relatives or neighbours who take advantage of social turmoil and government weakness to deny their claims to land.[2] In some cases, returning women will lose access to land because prevailing social or legal norms mediate their entitlement to land through a deceased or missing husband or relative. This is particularly the case for women who are widows, or who stand to inherit land from a deceased relative.

While these gendered risks of dispossession after conflict or disaster are real enough, the appropriate form and focus of any response is not so clear-cut. For humanitarian actors such as UN agencies and international NGOs, the orthodox prescription is to "mainstream" gender into rights-based programming. For example, a 1998 resolution of the UN Sub-Commission on the Prevention of Discrimination and Protection of Minorities notes the impact of land-related discrimination on women who are internally displaced. It urges governments[3]

> to take all necessary measures in order to amend and/or repeal laws and policies pertaining to land, property and housing which deny women security of tenure and equal access and rights to land, property and housing, to encourage the transformation of customs and traditions which deny women security of tenure and equal access and rights to land, property and housing, and to adopt and enforce legislation which protects and promotes women's rights to own, inherit, lease or rent land, property and housing ...

To similar effect, the 2005 United Nations Principles on Housing and Property Restitution for Refugees and Displaced Persons states:[4]

> States should ensure that housing, land and property restitution programmes, policies and practices recognize the joint ownership rights of both male and female heads of the household as an explicit component of the restitution process, and that restitution programmes, policies and practices reflect a gender-sensitive approach.

These types of policy prescriptions emphasize law and enforceable legal rights as central responses to gender risks after cases of disaster or conflict-induced displacement. "Customs and traditions" are considered in a negative sense only, as phenomena that require transformation of their discriminatory elements. The ensuing conception of law is both instrumentalist and centralist in nature. It assumes that law and legal institutions can act as direct instruments of desired social change, and it privileges the state as the provider of law and institutions that respond to gender-related risks of dispossession.

Land policy prescriptions based on law and legal rights must take into account the distinctive social relations and power structures that shape local claims to land. Local arrangements that receive the label "customs and traditions" are notoriously resilient in the face of attempts at legal transformation. They often continue in the shadow of law, creating uncertainty if disputants "shop" their claims to differing sources of legal and normative legitimacy. And while local arrangements can dispossess women of legitimate rights to land when changed circumstances favour claims by local power elites, they can also produce positive negotiated solutions for women left without access to land after conflict or disaster.

Based on extensive fieldwork after the tsunami in the Indonesian province of Aceh, this chapter argues that efforts to protect women's rights to land through formal law and legal rights produced a number of unanticipated and counter-intuitive results for Acehnese women, particularly in terms of negotiated and contested processes at the local level. For example, the primary formal response to land rights uncertainty after the tsunami was a land titling programme that formalized land rights through community-based agreements. This programme — the Reconstruction of Land Administration Systems in Aceh and Nias (RALAS) project — addressed the well-documented risks to women from land-titling projects by including a large number of gender safeguards. Yet our field data indicate that relatively few women were recorded as sole or joint owners of their land parcels. Indeed, it appears that women may have been relatively worse off after RALAS land-titling than under pre-existing localized arrangements.

The RALAS project also made provision for legal heirs to be recorded as landowners, after receiving inheritance approval from the village head (*keucik*) or village imam, and then from a mobile Syariah Court.[5] With at least 150,000 Acehnese killed by the tsunami, these decentralized processes were essential to restoring land-right certainty and to rehousing surviving family members after the disaster. Yet as a response to gendered risks of dispossession, the field data suggest that allowing more legal space for locally negotiated solutions that applied forms of custom rather than some provisions of Syariah law would have produced more favourable results for at least some female land claimants.

This is not to say that all local arrangements produce more favourable results to women than formal law and legal rights. In a number of cases, village leaders were unable or unwilling to prevent male relatives from denying legitimate inheritance claims by widows or daughters. But the field data do suggest that humanitarian actors should respond to gender risks by focusing more on local processes for asserting and negotiating claims to land and avoiding overreliance on top-down attempts at formalizing land claims, even though those formalization attempts include substantial gender safeguards. In particular, it appears that the provision of information, for supporting the assertion and negotiation of claims by women, should be a key focus of localized policy interventions.

The data presented in this chapter are based on interviews of:

- Women in Aceh Besar, Banda Aceh and Aceh Jaya
- Village heads and village leaders in Aceh Besar, Banda Aceh, Simeulue, Aceh Barat and Aceh Jaya
- NGOs active in community land-mapping and women's land rights
- Members of the National Land Agency's titling adjudication teams
- Judges of the Syariah Court in Banda Aceh and Jantho.

I have also drawn on field material kindly provided by the International Development Law Organization (IDLO). This field material dates from early 2006, and in some cases my research staff re-interviewed IDLO informants to update developments.[6]

Formalizing Land Titles after Disasters: The RALAS Project in Aceh

In Aceh, the primary formal response to land rights uncertainty after the disaster was the Reconstruction of Land Administration Systems in Aceh and Nias ("RALAS") project. The RALAS project was proposed by the Indonesian government in early 2005, and designed by a World Bank team in March 2005.

The core element of RALAS was a programme of systematic land title certification based on community-driven adjudication of land rights. Local survivors agreed on land ownership and boundaries through a process of community land-mapping or Community Driven Adjudication (CDA). The IDLO neatly summarized the steps involved:[7]

1. Each landowner must install boundary stakes and complete a statement attesting to the location of, and their ownership over, a specific land parcel. This statement must be endorsed by the owners of neighboring land and the village head (*keucik*).
2. Where the landowner is deceased, this form should be completed by the deceased's legal heirs who have previously received inheritance approval from their *keucik* or village imam (*imam meunasah*) (annexure 2, RALAS Manual).
3. Where heirs are minors, the form should be completed by a guardian approved by the *keucik* or *imam meunasah* and confirmed by the Syariah Court (*Mahkamah Syar'iyah*). As part of the RALAS programme,

the *Mahkamah Syar'iyah* will come to each village and conduct such confirmation hearings free of charge (annexure 3, RALAS Manual).
4. From these statements, communities then develop a map identifying the ownership and boundaries of land parcels in the village.[8]

The 2005 RALAS Manual further provides that, once the statements of ownership are complete, the *Badan Pertanahan Nasional* (the Indonesian National Land Authority, or BPN) is to survey the boundaries of identified land parcels using its accredited surveyors. After completion of the survey, BPN prepares a community land map that identifies boundaries and owners. This map is displayed on the village notice board for thirty days, in which time objections may be lodged for consideration by a village meeting or a provincial BPN complaints team. Subject to the determination of objections, BPN will issue land certificates to landowners within ninety days of the commencement of survey work.[9]

The 2005 RALAS Manual includes a large number of safeguards against the well-known gender risks of land-titling programmes. The most important are extracted in Part 2 of the manual, under the heading "How is the protection of women's rights over land and to other vulnerable groups conducted?". The key provisions of Part 2 are as follows:

1.1 Every right holder over land, both male and female, can register her/his land and obtain their certificate of ownership in her/his name.
1.2 A land certificate can be made in the name of more than one person. For land that is the joint property of husband and wife, the certificate can be made jointly in the names of other persons, and not just the husband.
1.3 For the inheritance of land whose share of entitlements are still not settled at the time adjudication is made, the certificate can be made in the names of a number of heirs, including those who are female and are still underaged.
1.4 According to the Syariah law, widows and female children have inheritance rights to land. Therefore, widows and female children who obtain inheritance rights over land must register their rights in their respective names.
1.5 Women landowners are expected to be present in the village meeting to discuss the plan for the village's land adjudication because it also concerns their interests.
1.6 Landowners (both male and female) must write their statement letter claiming the ownership of the land. They must also install boundary states themselves, unless they are declared incapable to do so.

1.7 Landowners (both male and female) must be present at the land parcel when the BPN adjudication team attends to undertake the adjudication, surveying and mapping.

1.8 For the distribution of certificates, the rightful owner (including women) must obtain the certificate themselves. Therefore, the adjudication team must ensure that women and widows can be present in the stage of certificate presentation, including the provision of any necessary security.

Recording Women's Rights to Land under RALAS

Partial data obtained from community-mapping and land-titling teams suggest that these gender safeguards had relatively little effect in practice. It seems that most land parcels were recorded in the name of men. In the districts of Banda Aceh and Aceh Besar, around 20–25 per cent were recorded in the name of women. Very few were recorded in the names of both husbands and wives.

The following data sets illustrate these findings. We begin with preliminary RALAS data for Banda Aceh and Aceh Besar, then turn to two villages in Banda Aceh (Lambung and Ulee Lheue). These data sets were obtained between July and November 2006.

These figures were obtained from RALAS staff. RALAS has not developed comprehensive gender-disaggregated data of its own. While the relatively low figures for women-only certification is a cause for comment, the most

TABLE 6.1
Preliminary RALAS Data for Banda Aceh and Aceh Besar

Distribution of Land Title Certificates by Gender in Banda Aceh and Aceh Besar						
Type of Certificate	Banda Aceh	% of District Total	Aceh Besar	% of District Total	Total	% of Total
Male name only	4163	68.2	1425	69.2	5588	68.4
Female name only	1699	27.8	561	27.3	2260	27.7
Male and female names	246	4.0	72	3.5	318	3.9
Total	6108	100.0	2058	100.0	8166	100.0

TABLE 6.2
Desa Lambung RALAS Analysis

Distribution of Land Titles by Gender in Desa Lambung		
Gender Breakdown of Certificate	Number of Land Parcels	% of Total
Male name	215	60.7
Female name	72	20.3
Co-owned including female	28	7.9
Co-owned males only	39	11.0
Total	354	100.0

TABLE 6.3
Ulee Lheue RALAS Analysis

Distribution of Land Titles by Gender in Ulee Lheue		
Gender Breakdown of Certificate	Number of Land Parcels	% of Total
Male name	275	81.8
Female name	48	14.3
Co-owned including female	8	2.4
Co-owned males only	5	1.5
Total	336	100.0

notable results concern the relative lack of joint titling of marital land. The available data suggest a very low proportion of co-owned land where one recorded owner was female, with results ranging from Lambung (7.9%) to Ulee Lheue (2.4%).

While it is not possible to make definitive judgements without further data, which the RALAS project has not made publicly available, it seems unlikely that the very small proportions of marital land recorded through joint titling in the RALAS project reflect the actual amount of marital land in tsunami-affected areas. This is particularly the case for urban land in Banda Aceh (including Lambung and Ulee Lheue). Under Indonesian law and Acehnese custom, land acquired in the course of marriage is co-owned by husband and wife. While rural land used by a married couple is usually land inherited either by the wife or husband — and hence individually owned — greater rates of labour mobility and land scarcity in urban areas make it more likely that married couples will have acquired land for their residential purposes. In other words, it is likely that the rates of co-owned marital land in Banda Aceh exceed the very low proportions recorded through the RALAS process.

Despite substantial gender safeguards, there is evidence to suggest that the RALAS project formalized land rights in a manner that did not record all the names of female landowners. The reasons appear to be both cultural and institutional in nature. A number of women interviewed for this field study indicated that their husbands would always be the house representative for formal documentation purposes, including in relation to land title certificates, even though the land was co-owned in terms of local norms and practice. It also appears that land agency staff facilitating land-title documentation did not ensure that ownership of land identified as marital property was recorded in the names of both the husband and wife. That this is so is not surprising, considering the widespread view that BPN lacked the capacity to implement the ambitious goals of the RALAS project.

The Aceh Institute has published a breakdown of RALAS data for the districts of Pidie, Bireun and Banda Aceh.[10] These data simply record the total number of men and women in the registration lists for each sampled subdistrict. They do not identify the number of co-owned land parcels, or the proportions held by women in a co-owned parcel. In the municipality of Sigli, Pidie District, there were 448 women (52%) and 414 men (48%) in the registration lists. These figures reflect the uxorilocal and daughter-inheritance traditions of Pidie. In two villages in Bireun, there were 347 women (40%) and 510 men (60%) recorded in the registration lists. In four sampled villagers in Banda Aceh, there were 686 women (32%) and 1,443 men (68%) in the registration lists. The Aceh Institute noted that joint titling of marital land is "extremely rare".[11]

In December 2005, a comprehensive survey of 347,775 out of a total of over 500,000 displaced persons in Aceh was conducted in Aceh by the NGO *Garansi* and the Indonesian Bureau of Statistics (the BPDE survey).[12] In this survey, 44.8 per cent of all displaced women stated that they owned land. While this figure may be reflected in the RALAS data sets for Pidie (52%) and Bireun (40%), it is not reflected in the preliminary available data for Banda Aceh (27.8%) and Aceh Besar (27.3%), including the Banda Aceh villages of Lambung (20.3%) and Ulee Lheue (14.3%). Hence, there are also indications that the RALAS project did not fully record the names of women who owned land in their own right. It is unfortunate that this preliminary finding cannot be confirmed by reference to comprehensive gender-disaggregated data published by the RALAS project itself.

THE INHERITANCE OF WOMEN'S RIGHTS TO LAND IN TSUNAMI-AFFECTED ACEH

Family Consensus and Locally Negotiated Solutions

The quantitative evidence that gender safeguards in the RALAS project were not fully effective in practice is supported by substantial qualitative material collected by my research team. Under the RALAS project, legal heirs were recorded as landowners once they received inheritance approval from their *keucik* or village imam. The transfer of the inheritance right would then be confirmed by a mobile Syariah Court. But the field data suggest that the inheritance of land in Aceh was determined primarily at the local level, without a great deal of reference to mechanisms established by the RALAS project. Most cases were resolved consensually among family survivors. Even village heads were unaware of many inheritance-based land transfers, even though they or the village imam were supposed to provide approval under the RALAS project. Most of these inheritance determinations also took place well before RALAS staff arrived to facilitate the recording of land ownership.

While Syariah law has formal applicability to inheritance determinations in Aceh, it appears common for family groups and village leaders to shape solutions for inheritance cases that are more beneficial to women than the strict provisions of Syariah law. Many of our interviewees noted that the Syariah inheritance rules prescribe 2:1 in favour of sons over daughters, but that in practice it is very common for families to agree to a more equitable division.[13]

In an interview in 2006, the Chief Justice of the Syariah Court in Aceh stated that the mobile Syariah Court would verify inheritance allocations agreed by all family members, even though the proportions of Syariah law had not been applied. Nevertheless, our fieldwork identified a high degree of uncertainty among *keucik* and village imams as to their authority to shape inheritance solutions for women that did not comply with the proportions mandated by Syariah law.[14] This finding supports the conclusion that localized information flows are critical to the success of policy responses to gendered risks of land dispossession after conflict or disaster.

BOX 6.1
Examples of Locally Negotiated Inheritance Solutions in Tsunami-Affected Aceh

Grant of Land to Widows
Wardiah and her husband Meka did not own land prior to the tsunami. Instead, they worked for someone in the village who owned fishing ponds and they eventually built a house on that land. That house was destroyed by the tsunami. Meka and all of their children also passed away. Wardiah was thus a landless widow with no children. Uplink, an NGO, offered Wardiah housing assistance. A villager granted Wardiah a piece of land on which Uplink has now built a house.

Inheritance of Land by Female Orphans
Lia was orphaned by the tsunami. Generally, a female orphan will inherit half of her parents' estate with the balance going to a wali (guardian) from the orphan's paternal lineage. In a breakthrough decision, the keucik of Kahju granted Lia her parents' entire estate. The decision to allow the daughter to inherit the entire estate was based on the deliberations of public figures in Kahju. Relevant factors included the fact that the girl was seventeen years old and still at school, and that no other persons contested the decision. The estate was also being managed entirely by the orphan as no guardianship arrangement was in place.

Barniah was orphaned by the tsunami. According to Islamic law as practiced in her community, as a single daughter Barniah is entitled only to half of her parents' inheritance. The balance is given to her guardian, who in this case is Barniah's uncle, Zulhamsyah. Through his own benevolence, Zulhamsyah chose not to receive inherit-

---

The Application of Custom in Aceh

Our fieldwork also confirmed evidence in the literature that local custom and practice relating to women's rights to land apply in certain parts of Aceh. For example:

The uxorilocal tradition. The custom in some Acehnese districts (including Pidie and parts of Aceh Besar) is that daughters receive land (and perhaps a house) from their parents at the time of their marriage. Traditionally, this land will be in the same compound as the parents' house. The daughter (wife) is recognized as the sole owner of the land (and house).

The inheriting daughters tradition. In parts of Pidie, Aceh Besar and even Banda Aceh itself, daughters will inherit their parents' house and surrounding residential land. Their brothers will inherit non-residential land. In some

---

BOX 6.2
Uxorilocal Traditions in Aceh

Women interviewees in Pekanbada, Aceh Besar, stated that the uxorilocal tradition operates in their village. Usually the wife's parents provide their daughter and son-in-law with land and a house. Some will provide only land. In cases where parents cannot provide land or housing, the daughter and son-in-law will live in the wife's family home until such time as sufficient funds can be raised for the husband and wife to build a new house. One woman respondent said that, when she was married, her mother granted her land on which she and her husband consequently built a house. Prior to the tsunami, there was a land certificate for this piece of land in her mother's name. Her mother died in the tsunami and, during the land adjudication processes, BPN registered the land in the woman's name.

BOX 6.3
Inheritance Rights for Youngest Daughters in Parts of Aceh

Inheriting daughters in Banda Aceh
Gampong Teungoh, Kec. Meuraxa, Banda Aceh applies a mix of adat (custom) and Islamic law in determining inheritance. Adat prescribes that sons are entitled to agricultural land, while daughters inherit residential land and housing. The youngest daughter, who has responsibility for caring for her parents' health, has priority to inherit residential land and housing. Once inherited, the daughter has full ownership of the land and can bequeath, gift or sell the land; however, any proposed sale must be discussed with relatives who have a right of first refusal.

---

places, the youngest daughter only will inherit her parents' residential land and house. In return, the daughters (or youngest daughter) are expected to care for her parents up until their deaths.

Surviving spouses. Syahrizal, an expert on Acehnese custom, also suggests that Acehnese custom in certain areas may recognize the inheritance entitlement of surviving spouses to a greater extent than Syariah law.[15]

There is a high degree of ambivalence among international actors towards custom and its relationship with women's legal rights after conflict or disaster. Our field evidence suggests that localized processes that have been institutionalized as custom should not necessarily be viewed as inherently discriminatory and antithetical to the rights of women. While there is no evidence that the RALAS project and the mobile Syariah Court acted to override local customs in favour of Syariah law, there is a case for policy-makers to have focused more on the potential benefits of custom in particular areas of Aceh.

Denial of Women's Claims: Village Leaders and Male Relatives

Local processes do not necessarily favour women. There was anecdotal evidence of local leaders denying widows their legitimate claims to land, even when these plans were supported by Syariah law. These anecdotal accounts include pressure on widows to remarry in order to claim their land. Our research did not reveal any direct evidence of dispossession of women's rights to land by local officials that was outside the framework of Syariah law. This research included meetings with leading women's NGOs in Aceh, as well as a large number of interviews with women in Banda Aceh and Aceh Besar.

Our research did reveal a number of cases where village leaders were unable or unwilling to prevent relatives from denying legitimate claims by widows or daughters. These dispossessory acts by relatives built on the social powerlessness felt by many women after the tsunami. They took the form of:

- Implicit threats of violence by male relatives
- Arguments that female claimants could not obtain land unless they married (or remarried)
- Arguments that widows could not claim land because the land had been independently owned by their husband (i.e., it was not marital land).

While village heads have legal authority to resolve these village-level disputes under Regional Regulation 7/2000, a number appeared to be unwilling or unable to resolve family conflicts that involved the denial of claims by women.

Too much reliance was placed on gender safeguards in the RALAS project to avoid these local abuses of power in inheritance matters. Most inheritance determinations antithetical to women occurred at the local level, without connection or reference to the RALAS project; yet there was very little policy or programming attention paid to this possibility apart from information campaigns supported by the Syariah Court, Oxfam and IDLO.[16] The next section's findings on information availability suggest — at least in gender protection terms — that international funding for the

BOX 6.4
Localised Denial of Women's Rights to Land in Parts of Aceh

A Violent Brother-in-Law
Witni works as an official for an Indonesian company that distributes mobile phones. Prior to the tsunami, her husband was unwell for a long period of time and Witni's wages were the sole family income. Witni is now a widow—she survived the tsunami but her husband and children did not. Witni is extremely traumatized by the disaster and, worse still, is now being deprived of her inheritance by her husband's four siblings who survived the tsunami. Despite Witni's legal entitlements, her husband's siblings have claimed 100 per cent of her husband's estate. Witni has been left with nothing and has no confidence to protest to her husband's eldest brother about the inheritance, as he has a reputation for violent behaviour. Witni feels alone and without support.

Does a Granddaughter Need to Marry in Order to Obtain Land?
Salwa passed away three months prior to the tsunami, leaving four sons and three daughters (Salwa's wife had predeceased him). Salwa's estate consisted of one hectare of land. Only one son of Salwa's (Kautsar) survived the tsunami. A sister-in-law named Lily also survived. While Lily's husband died in the tsunami, her daughter Jeni survived. According to Indonesia's Compilation of Islamic Law, Jeni should receive a portion of Salwa's (her grandfather's) estate even though her father passed away in the tsunami. However, Kautsar does not want Jeni to receive the inheritance now. Kautsar claims that Jeni is only entitled to rights in her father's land upon marriage. The matter remains unresolved.

Marital Land or Husband's Land?
A women named Rasmiah survived the tsunami, but her husband Zulfan did not. Prior to the tsunami, Rasmiah and Zulfan had jointly owned land. Rasmiah had a house built on this land by an NGO after the tsunami; however, following the construction of the house, Zulfan's family attempted to claim ownership of it on the basis that Rasmiah was not entitled to the land. It is not clear whether this matter has been settled.

Children's or Brother's land?
In Peukan Bada, Aceh Besar, two orphans were denied the right to inherit their mother's land and have a house built on it by an aid agency. The mother's land originally belonged to the orphans' grandmother, who had passed it to their mother prior to her death without a letter of bequest. After the tsunami, the land was taken over by the orphans' uncle (the mother's younger brother). Community leaders considered the orphans ineligible to inherit the land and supported the uncle's takeover, notwithstanding that the children are legally entitled to receive that land and have a house built on it.

continued on next page

BOX 6.4 — *cont'd*

Wife and Mother, or Brother?
Yusnita is from Aceh Besar. Yusnita lost her husband and children in the tsunami. Despite
having a legal right to the large amount of inheritance left by her family (including
land), Yusnita received nothing—the estate is now controlled by Yusnita's husband's
elder brother. When Yusnita reported the case to the keucik, she was ignored. When
she approached her husband's brother, she was threatened.

Granddaughters Denied Their Right to Inherit
Tiramah passed away in the tsunami. Village authorities (the keucik and the village
secretary), insisted that Tiramah's granddaughter Ayusnita is not eligible to inherit
Tiramah's land, because Ayusnita's mother had passed away before the tsunami (patah
titi). The tengku (religious leader) stated that Ayusnita was eligible. Another one of
Tiramah's granddaughters was killed by the tsunami and was not allowed to be buried
in the family graveyard because the child was no longer considered the grandmother's
descendant.

RALAS project may have been better spent on extending and deepening
these information campaigns.

Information Needs: Village Leaders

In our field interviews, a number of village heads and imams admitted
to uncertainty concerning the proper application of Syariah law to more
complex cases. Complex cases are relatively widespread because the tsunami
created every conceivable kind of inheritance possibility. These complex
cases often remain unresolved, or at least uneasily settled and susceptible to
future conflict, because (1) village leaders may not be willing or able to give
an authoritative determination, (2) the claimants refuse to accept a decision
by the village head, or (3) the claimants may not be willing or able to bring
the case to the Syariah Court.

In our field interviews, some village leaders and priests expressed
uncertainty as to the applicable law in the following types of cases:

o   The rights of female orphans
o   Distinctions between joint and independent property
o   The rights of residual heirs when an entire family had died
o   The relative rights of daughters and uncles

o   The rights of widows, particularly as against brothers of their deceased husband
o   The relative rights of granddaughters and sons
o   The status of marital property after divorce when there has been no divorce settlement.

All these cases involved uncertainty as to claims by women.

BOX 6.5
Local Uncertainty in Complex Inheritance Cases

The Relative Rights of Nieces and Brothers
Numeri did not survive the tsunami. Three survivors laid claim to his estate—two brothers and a niece, whose father (Numeri's brother) also died in the tsunami. The issue that emerged was whether the niece was entitled to inherit. Numeri's surviving brothers contested her claim. In their opinion, the niece was disentitled to a share of the estate when her father passed away. The case was reported to the keucik, who was not sure how to resolve it.

Information Needs for Women

Our fieldwork involved a large number of group and individual interviews with women in Banda Aceh, Aceh Besar and Aceh Jaya. In these interviews, women expressed a degree of confusion and a desire for further information concerning:

• The rights of widows to both independent and jointly acquired property
• The right of orphaned grandchildren to inherit from their grandparents (*patah titi*).

For example, women respondents from Lhok Seude were not aware of the formula used to divide property for widows.[17] Women respondents from Kahju barracks were aware that widows have the right to inherit marital property, but were not sure whether widows have the right to inherit their husband's independent property.[18] The Kahju women also had differing opinions on how marital property would be divided upon divorce or death.

Some respondents stated that property would be halved between husband and wife, while others stated it would be divided into three between husband, wife and their children.[19] In Ulee Lheue, women respondents had mixed understandings of their rights over their husband's independent property. One stated that if there is no child from the marriage, a wife is not entitled to inherit her husband's independent property. Another said that she is not entitled to inherit her deceased husband's independent property as her husband has children from a prior marriage. A third said that she could inherit her husband's independent property on the basis that all of his property would be divided into two.[20]

A number of interviewees argued that access to information is the best mechanism to ensure women's rights and access to land.[21] Both the Syariah Court, with the assistance of Oxfam, and IDLO undertook information campaigns that include these and other inheritance issues. As noted, these programmes deserved much more international funding to support the negotiation and assertion of claims at the local level.

CONCLUSION

Relationships with land are socially embedded. Local norms govern the use of land. Localized interpretations affect the scope and enforceability of claims to land. Even in circumstances where the state is strong and access to legal institutions readily available, there will still be a high degree of uncertainty or conflict if a claim to land is not recognized by neighbours or other claimants. And if the state is weak or widely distrusted, efforts to impose law and legal rights over pre-existing local arrangements may only create unintended forms of legal or normative pluralism.

Women's rights to land after the tsunami were not simply an issue for international actors, arising from international legal frameworks. They were a major source of contestation and negotiation at the local level. International instruments that focus on programmes to restore land after displacement in the names of men and women, or that contemplate the transformation of customs and traditions to rid them of their discriminatory elements, are inadequate in themselves to prevent unwarranted post-denial of women's land rights if they fail to take into account the fact that local processes fundamentally affect the way in which law and legal rights are mediated in practice. The point is not to say that local processes are either good or bad for female land claimants. It is to highlight the importance of local processes in post-disaster programming, and to question the utility of overreliance on gender safeguards in land rights formalization programmes.

Notes

1. For examples and discussion, see The United Nations Development Fund for Women (UNIFEM), "Women's Land and Property Rights in Situations of Conflict and Reconstruction" (2001).

2. United Nations Centre for Human Settlements (UNCHS (UN-Habitat)), "Woman's Rights to Land, Housing and Property in Post-conflict Situations and During Reconstruction: A Global Overview" (1999) 2–3.

3. UN Sub-Commission on the Prevention of Discrimination and Protection of Minorities, Res 15 (1998), "Women and the Right to Land, Property and Adequate Housing', UN doc E/CN.4/Sub.2/RES/1998/15.

4. United Nations Economic and Social Council (ECOSOC), "Housing and Property Restitution in the Context of the Return of Refugees and Internally Displaced Persons, Final Report of the Special Rapporteur, Paulo Sérgio Pinheiro, Principles on Housing and Property Restitution for Refugees and Displaced Persons" (28 June 2005) UN Doc E/CN.4/Sub.2/2005/17.

5. The mobile Syariah Court involved judges accompanying or following RALAS adjudication teams to authorize uncontested inheritance decisions by surviving family members.

6. Funding for the field research was kindly provided by Oxfam UK.

7. See Harper, Fitzpatrick and Clark, *Land, Inheritance and Guardianship Law in Aceh* (Rome: International Development Law Organization, 2006).

8. Ibid., pp. 17, 24.

9. These timelines were rarely satisfied in practice.

10. Kadriah et al., *Perlindungan Terhadap Perempuan Korban Tsunami Dalam Mendapatkan Hak Kepemilikan Atas Tanah* [The protection of female victims of the tsunami in the context of land ownership], University of Syiah Kuala Darussalam Law School, Banda Aceh, December 2006.

11. Ibid., p. 21.

12. In doing so, the survey encompassed some districts holding displaced persons that were affected by the secessionist conflict in Aceh rather than the tsunami itself.

13. Interview with Zaenabun (from Meunasah Tuha Village, Kecamatan Pekan Bada), 28 November 2006; interview with Sudirman Arif (from Kade Gelumpang Teungoh, Desa Surien, Kecematan Meuraxa Banda Aceh), 27 November 2006; group discussion with Maemunah, Salmah, Aminah and Rifky (women from Meunasah Tuha, Kecamatan Pekan Bada), 28 November 2006; group discussion with eighteen women and one male village facilitator from LOGICA, Kahju barrack, Darussalam, Aceh Besar, 30 September 2006; interview with Zaenabun (from Meunasah Tuha Village, Kecamatan Pekan Bada), 28 November 2006; interview with Sudirman Arif (from Kades Gelumpang Teungoh, Desa Surien, Kecematan Meuraxa Banda Aceh), Uplink's Offices, 27 November 2006; group discussion with Maemunah, Salmah, Aminah and Rifky (women from Meunasah Tuha, Kecamatan Pekan Bada), 28 November 2006.

Between Custom and Law   131

14. To some extent, this uncertainty arises from longstanding attempts to integrate traditional Indonesian conceptions of joint marital property — defined as property acquired during the course of a marriage through the efforts of either spouse — into existing schools and doctrines of Syariah jurisprudence: for a discussion, see Mark E. Cammack and R. Michael Feener, "Joint Marital Property in Indonesian Customary, Islamic and National Law", in Peri Bearman, Wolfhart Heinrichs and Bernard Weiss, *The Law Applied: Contextualising the Islamic Shari'a* (London: I.B. Tauris, 2008), pp. 92–115.

15. See Harper et al., *Land, Inheritance and Guardianship Law in Aceh*, International Development Law Organization 2006, pp. 54–55.

16. See Part III C below.

17. Group discussion with four women (one aged widow, two middle-aged ladies and one young lady), Lhok Seude, Leupueng, Aceh Besar, date unknown.

18. Group discussion with eighteen women and one male village facilitator from LOGICA, Kahju barrack, Darussalam, Aceh Besar, 30 September 2006.

19. Group discussion with eighteen women and one male village facilitator from LOGICA, Kahju barrack, Darussalam, Aceh Besar, 30 September 2006.

20. Group discussion with women from Dusun Tenggiri, Desa Ulhe Lhe, Meuraxa Banda Aceh, 28 November 2006.

21. Interview with Yatrin (LOGICA Gender Specialist) and Sofyan (LOGICA Community Land Mapping Specialist), LOGICA Offices, 29 September 2006.

7

FACTORS DETERMINING THE MOVEMENTS OF INTERNALLY DISPLACED PERSONS (IDPs) IN ACEH

Saiful Mahdi

INTRODUCTION

"Migration from an area afflicted by a major disaster to an unaffected area would seem to be one of the most common responses to disaster and an important survival strategy".[1]

Aceh has witnessed significant human migration due to both man-made and natural disasters. The civil war during the late 1940s to early 1950s, followed by three decades of struggle for independence since 1976 have caused major IDP and refugee crises. The great quake and tsunami of 26 December 2004 caused another major IDP crisis, displacing more than half a million people.[2] It is widely established that disasters and conflicts typically influence population movements, as well as the social characteristics of affected communities.[3] Natural disasters and conflicts often generate both large- and small-scale migrations of people away from affected areas.[4] Paul, however, argued that not all affected communities out-migrate permanently after a disaster, if there is a "constant flow of disaster aid and its proper distribution by the government and non-governmental organizations (NGOs)".[5] This chapter uses the case of IDPs in Aceh to look in more detail at the impact of aid distribution upon mobility and community cohesion in post-disaster situations.

In my research, I take the position that migration was actively used as a mode of survival for the Acehnese during the conflict and initial days after the tsunami. I relate this to "social capital" and the importance of village (or *gampong*) connections in Aceh. I demonstrate that social capital is embedded within *gampongs,* and was pivotal to both the movement of conflict and tsunami IDPs. In this chapter, I demonstrate that the main force determining the movements of IDPs from the long-running conflict between GAM and the Indonesian military were village connections and networks. This provides a control variable for looking at the impact of post-tsunami relief and reconstruction aid on IDP responses. I argue that *gampong* connections were undermined after the tsunami by relief- and reconstruction-related interventions. Finally, I draw on two contrasting micro-studies on how communities adapted post tsunami, to discuss the impact of outside interventions on the level of success of reconstruction projects.

SOCIAL CAPITAL AND THE ACEHNESE *GAMPONG*

I see Acehnese *gampong* connections as networks built on trust with specific norms and values regulating them, fitting to a notion of "social capital" as prescribed by various social scientists.[6] Bourdieu noted that "social capital is the sum of resources, actual or virtual, that accrue to an individual or a group by virtue of possessing a durable network of more or less institutionalized relationships of mutual acquaintance and recognition".[7] Putnam linked "the civic community" to the tradition of "civic humanism", which he specifies has four theoretical dimensions: (1) civic engagement, (2) political equality, (3) solidarity, trust and tolerance, and (4) the social structure of cooperation. Putnam found in his study that northern regions in Italy performed better than those of southern ones, as "social capital" flourished in the former:

> Collaboration, mutual assistance, civic obligation, and even trust ... were the distinguishing features in the North. The chief virtue in the South, by contrast, was imposition of hierarchy and order on latent anarchy.[8]

Fatimah Castillo examined cultural adaptation and social capital aspects of resilience.[9] Castillo argues that, in many cases, the fabric of social life and social capital — friendship, kinship, trust, religious beliefs and indigenous belief systems — are resilient in the face of conflict and disaster. Therefore, "just because many have died, or that physical infrastructure is destroyed", this does not always mean "that the fabric of social life is also destroyed". In this section, I want to discuss some of the basic mechanisms of the Acehnese *gampong* that are relevant to understanding Acehnese capacities to respond to IDP situations. This first involves a brief description of *gampong*, as well as a discussion about how forms of "resilience" and response are part of local social fabrics.

GAMPONG AND ACEHNESE SOCIAL CAPITAL

The *gampong* is "the smallest territorial unit" in Aceh and elsewhere within Indonesia.[10] *Gampongs* were initially the product of *kawoms* (kinship) or of their subdivision, which added to their number only by marriage from within or at most with women of neighbouring fellow tribesmen. Traditionally, a *gampong* was headed by a *keucik*[11] (village head), who used to be chosen from among *uleebalangs* (chiefs) of the *kawom* or their successors. Later, however, as *kawoms* got bigger and boundaries began to overlap, *keuciks* were chosen more democratically. According to Qanun No. 5/2003, a *keucik* is the executive of the *gampong* in day-to-day governance. The *keucik* works together with *tuha peut* (village elders), *teungku* or *imeum meunasah* and the village secretary to provide basic governance of the *gampong*.

The *keucik* is not only a leader of his people and territory, but also the caretaker of *adat,* or customary law. The *keucik* is elected democratically by people in each *gampong* while the more modern *lurah* or *kepala desa* (village head) are appointed in a top-down fashion by the government through the *camat* (head of *kecamatan,* subdistrict). Additionally, all matters of public interest in a *gampong* are discussed openly and decisions are made based on *mupakat* (derived from the Arabic *muwafakat*). Life in a *gampong* is very communal, with strong association among its members. Relations among members of a *gampong* are not based on self-interest, but more on common understandings of harmony, accordance and appropriateness, as shown in Table 7.1.[12]

HUKUM AND *ADAT GAMPONG* AS THE BASIS
FOR RESILIENCE

The close and obvious relation between *hukum* (religious law) and *adat* in Acehnese communities does not mean that the two are always the same or in agreement. Snouck noted that

> a distinction is drawn in practice between what is religious in the strict sense, and therefore inviolable and what is of more secular nature and may accordingly be modified to suit the requirements of the state and of the society, or even altogether set aside. This explains the contrast which the Achehnese express by the words *huköm* and *adat*.[13]

TABLE 7.1
Community Formation and Association in Acehnese Society

Community	Form of Association	Remark
Genealogic	*Kawom-syedara* (kinship)	Based on parents' track, following both mother and father's side
Territorial *gampong*	*Syedara lingka* (territorial relation)	Starting from neighbour around one's house to the whole
Genealogic — Territorial	*Syedara gampung* (territorial-kinship)	It is common to find that inhabitants in a *gampung* are related to one another

Source: Tripa (2006)

Still, all laws alike possess religious character, including those norms and values that have to do with resilience and resistance. Siapno reported that Acehnese people

> gave at least three reasons for their resiliency in times of conflict and displacement: (1) the tradition and structure of experience of *perantauan* (the non-original place — being on a journey), (2) *tawakkal* (submission to and trust in God who orders everything); and (3) *saudara seperjuangan* (kindred spirit networks, communities of displaced persons).[14]

All these reasons, especially the first and third, are directly related to the concept of *silaturrahim* (or *silaturrahmi*, bond of good friendship or brotherhood) and nurtured by Islamic values confined strongly, although not exclusively, in *gampongs*. Additionally, the two concepts cannot be separated from the notion of mobility and, thus, geographical setting. This gives "space" to what Siapno pictures as more of a mental state. I argue that strength and resilience are not as strong in individuals in Aceh as they are in the collective *gampong*.

When Acehnese face hardship, many people fall back on *tawakkal* and *silaturrahim*. *Tawakkal* is more personal, where one turns to one's belief in God for everything that happens. One should therefore not become engrossed in prolonged sadness, as everything is God's will. *Silaturrahim,* in contrast to *tawakkal,* is a community virtue, where people strengthen one another in calamity. It is a religious and customary obligation to visit and soothe members of communities who are distraught. It is religious because it is held that all Muslims are brothers and sisters, as prescribed by one of the sayings of the Prophet Muhammad (*hadith*): "Muslims are like a body, when one part feels pain, another part will also feel the pain". It is customary because in *gampongs* people know one another and often visit one another. The visits are more essential during important times of one's life: births, marriages and deaths. If one excludes oneself from these exchanges of visits, one might lose one's rights for reciprocation. *Perantauan,* is according to Siapno

> the mentality of always being on a journey ... a structure of experience which is not prone to parochialism but towards open-mindedness and constant striving. A tradition of constant striving, of constantly re-creating one's self, of living in uncertainty, of being an outsider in an established system — whether in search of *ilmu* (knowledge), vocation, or a job, as a precondition to a mature self in Aceh.[15]

Moreover, Siapno relates *perantauan* to the concept of *musafir* (travellers), which Islam identifies as being in the category of persons one should assist and show compassion to, besides *fakir miskin* (poor people) and *aneuk yatim* (orphans). The concept of *perantauan* and *merantau* (to travel, to migrate) dates back to early history, as Acehnese embraced the notion of *meudagang*, which literally means "to trade". Snouck (1906, p. 26) noted that "in Acheh, the word *meudagang*, which originally signifies 'to be a stranger, to travel from place to place', has passed directly from this meaning to that of 'to be engaged in study' ". In a footnote on the same page, he wrote "*ureung dagang* always means 'stranger' usually applied to traders ..." However, "*meudagang* has now no other meaning than that of 'to study' and *ureung meudagang* means 'a student' ". This concept of *meudagang* is still used and has always been in the psyche of the Acehnese. It is especially true among the people from Pidie, North Aceh, Gayo and South Aceh, who are famous for being great migrant traders and are fond of travelling to seek knowledge and better lives.

I have tried to focus on the aspects of life in Acehnese *gampongs* that form social capital and influence IDP mobility. It is clear that there is a strong *gampong* cohesiveness which is firmly grounded in kinship and family, and that *gampongs* play very central, practical and symbolic roles in Aceh. Furthermore, *gampongs* are equipped with established systems of leadership and local governance, headed by the *keucik* and village elders, which operate under a loosely democratic system. Finally, there are a number of traditions within Aceh that can potentially shape how the Acehnese view IDPs as parts of extended networks and as travellers. In many ways, the idea of displaced persons was relatively easy for the Acehnese to fit within the ways they maintain relationships, provide support and understand notions of mobility. As I will document below, during the conflict and early days after the tsunami, there were local responses to IDPs that not only gave direction to their movements, but also assured many of them some level of support and hospitality within *gampongs* that they had connections with. I will now move on to the practical issues of internally displaced people in Aceh as a result of the conflict and the tsunami, before turning to the case studies.

IDPs IN ACEH

Literature on human migration in Aceh is scarce. Reports on the situation of IDPs have been provided mostly by government agencies and human rights groups.[16] In terms of geographical boundaries, people who move across states are usually called *refugees*, while people who move within states are called *internally displaced persons* (IDPs). These terms are often used interchangeably; although, for political reasons, a country might insist on the use of the term IDPs instead of refugee. In Indonesia, for example, the term IDPs is used for people who move because of conflict and natural disaster; people who move both within or across borders are called *pengungsi* in Bahasa Indonesia, which has only this one word for both IDPs and refugees.

In 1999, *Inside Indonesia* reported that approximately 80,000 Acehnese were displaced from their rural villages in the three most conflict-prone districts (Pidie, North Aceh and East Aceh).[17] Hugo reported that, based on conditions at the beginning of 2002, Indonesia had one of the largest groups of IDPs of any nation in the world, with most located in the outer island provinces, including Aceh.[18] The global IDP Project, established by the Norwegian Refugee Council at the request of the United Nations to monitor conflict-induced internal displacement worldwide, reported that Indonesia (Aceh) has the world's tenth-worst displacement situation in 2003, which was well before the tsunami.[19] It is important to discuss the nature of conflict IDPs in Aceh, and to isolate some of the main factors determining their movement.

Conflict IDPs

Both GAM and the Indonesian military were responsible for displacements in Aceh during the conflict. During 1999–2000, GAM occasionally used IDPs and refugees to attract international attention:

> One tactic has been to cultivate the support of international human rights groups. Another approach — employed in mid-1999 — was to empty dozens of villages, and move between 80,000 and 100,000 Acehnese into 61 refugee [IDPs] camps, provoking a refugee [IDPs] crisis. After drawing

international media attention, villagers were allowed to return to their villages and these camps were largely closed down.[20]

Information compiled by ELSAM, a human rights NGO based in Jakarta, showed that during June–July 1999 there were at least 117,000 IDPs in Pidie, North Aceh and East Aceh.[21] The number decreased during the implementation of the Cessation of Hostilities Agreement (CoHA), brokered by the Switzerland-based Henry Dunant Centre for Humanitarian Dialogue, in 2000–02.

Following the abrupt halt of CoHA, Aceh was put under martial law on 19 May 2003, during which time the TNI (Indonesian army) used "forced evacuation" tactics to remove people from their villages and split them from the guerrillas, known as "separating fish from the water" in counter-insurgency warfare. An unofficial report of UN-OCHA (UN Office for the Coordination of Humanitarian Affairs) showed that there were 8,251 households, comprising 35,649 IDPs, between 19 May and 17 June 2002.[22] Two months after the imposition of martial law, IDPs increased to over 50,000 people spread in different regions throughout Aceh. Some 24 per cent of the total IDPs were children, women and the elderly.[23] By August 2003, according to International Organization of Migration (IOM), the total number of IDPs in Aceh reached 104,702 persons.

By October 2003, the IOM and its Indonesian counterpart, *Satkorlak*, the national authority for disaster mitigation, officially reported that 118,000 people were displaced by the military operation. Hedman, however, reported that

> These numbers only include IDPs in designated camps; many displaced have sought refuge with relatives or are in camps that are not managed by the martial law administration. For example, 3,780 persons from the Pasie Raja Sub-district in South Aceh sought refuge with host families. There have also been reliable reports of people fleeing into the forests. Furthermore, the figures do not account for displacement to other regions of Indonesia and asylum-seekers in other countries.[24]

Many people left or were forced from their villages in rural areas where hostilities were high, especially in the interior of Pidie, North Aceh, and East Aceh. In cases where mobility was possible, many IDPs usually moved to urban areas like Banda Aceh, Sigli and Lhokseumawe. For example, Kota Mini Beureuneun, a small city in Pidie and one of the hottest spots during the conflict, was much less vibrant as many traders moved away and settled in the Banda Aceh or Medan marketplaces. Many formerly abandoned and empty shops and warehouses in the Banda Aceh marketplaces were suddenly crowded by these IDPs. It has been noted that some IDPs moved out of Aceh to elsewhere in Indonesia as well as to other countries. It is well known that many Acehnese with sufficient means bought houses in Medan to escape the hostilities in Aceh. Additionally, other Acehnese sought refugee status in or through Malaysia to other countries.[25]

The movement of these IDPs was very much dependent on their connections to parties (family, friends, organizations, etc.) in those destinations. Based on observations from some of my fieldwork, family and friendship ties and "place of origin" (*gampong*) connections played major roles in these movements. Even the rebel groups who fled Aceh seemed to be bound more by this *gampong* concept than by their organizational structure.[26] In short, when people were forced to move because of the conflict, their default was either to move to one of the main urban centres, or to places where they had strong family or *gampong* connections (or in some cases a combination of both).

Tsunami IDPs

The situation of IDPs in Aceh after the tsunami proved that the definition of the term IDPs is not a trivial matter. A letter from United Nations Recovery Coordinator for Aceh and Nias (UNORC) to the Head of Aceh and Nias Rehabilitation and Reconstruction Agency (BRR) on 2 December 2005, about a year after the tsunami, illustrates this:

> Data on what UN Agencies and NGOs define as internally displaced persons or IDPs has represented a major problem in analyzing the post-tsunami humanitarian situation in Aceh and Nias. Credible data on these categories have been difficult to obtain due to the dispersed nature and mobility of affected communities and the varied arrangements of camp management, all preventing a uniform registration process of IDPs from occurring.[27]

Garansi, the contractor for local government's BPDE, defined IDPs as person(s) "being affected by the disaster and displaced from their residence and living in temporary accommodation".[28] There were reports that the *Garansi* survey in some places included host families as part of the IDP count. The SPAN census, on the other hand, defined IDPs based on whether respondents at the time of interview identified themselves as an IDP. Each respondent, typically the head of the household, was asked, "*Apakah saat ini Anda sebagai pengungsi?*" [Are you an internally displaced person at this time?]. This subjective and self-identified question might have resulted in a far lower count of IDPs than the numbers previously produced by other sources.[29]

IDPs from both conflict and natural disasters are often seen as victims with no or limited agency. In Aceh, however, being an IDP needs to be seen as part of strategy for resilience and resistance. Siapno noted two contrasting formulas for looking at "displacement" in general. On the one hand, there is the government's perspective, as seen in their reports, which often attribute no agency to "displaced masses". On the other hand, there is "human rights formulaic analysis", which tends to see the displaced as "the victims" and "the oppressed", missing "other subjectivities".[30]

When the question is put subjectively, as in the SPAN census, it might be surprising for some that "312,463 tsunami/earthquake survivors who once identified themselves as IDPs no longer do".[31] This number is the result from a follow-up question *"Apakah Anda pernah mengungsi setelah gempa/tsunami?"* [Have you ever been displaced as a result of the earthquake/tsunami?] This should not be surprising, because many communities decided to return to their original settlements as early as two to three months after the disaster. Many of these returnees were not recorded while they were actually "still living in tents or ad-hoc structures on their own land". Chris Morris, United Nations Recovery Coordinator for Aceh & Nias, sketched an explanation for these "ex-IDPs":

> There may also be cultural, psychological or linguistic reasons why such a large number of people who once considered themselves IDPs no longer want to identify themselves as such. Two possible explanations stand out. Firstly, many tsunami/earthquake survivors have gone back to their own lands and now consider themselves as having returned home and therefore no longer an IDP regardless of the living conditions they experience. Secondly, being labeled a "pengungsi" or IDP in Aceh may carry a negative connotation which many tsunami survivors probably want to avoid.

Aceh was still deeply involved in the conflict when it was further devastated by the earthquake and tsunami on 26 December 2004. This dramatically increased the number of IDPs in Aceh. In September 2005, there were around 436,820 IDPs in Aceh (not including Nias) in temporary living centres (TLC) such as government-built barracks (17.3 per cent), self set-up tents (15.5 per cent) and host communities (67.2 per cent). It is clear that host communities constitute the biggest "shelter" (67.2 per cent) for the tsunami IDPs in Aceh. People usually gravitated towards "host communities" where they had family members, relatives, friends or social relations based on *gampong* connections.

Many IDPs who had connections in communities in the interior went back to villages in the rural areas. This was especially true for people from

Pidie and Bireuen (northeastern), districts that were famous as sources of local migrants and traders to urban areas like Banda Aceh and Lhokseumawe. Although Pidie and Bireuen were not devastated as badly as Banda Aceh, Aceh Besar or Western Aceh, the communities in these two districts were greatly affected by the return of their people from the coastal urbanized areas. This phenomenon was indicated by the distribution of the number of tsunami IDPs in districts of Aceh, especially in the urbanized regions, as shown by Table 7.1.[32]

Numbers of IDPs in central Aceh regions like Aceh Tengah, Bener Meriah, Aceh Tenggara and Gayo Lues, which do not border the sea, indicate that people moved from coastal areas into the hinterland after the tsunami, with many probably returning to their original *gampong* from which they had been previously displaced (See Table in Appendix 2). People returned to their rural village *(gampong)* to mourn, to heal and to gather the strength and, in some cases, resources to start a new life.

In the above sections I have shown that one of the responses for IDPs in Aceh during the conflict and immediately after the tsunami was to move to areas where they had connections. These were mostly based upon informal networks of friends and family that made some *gampongs* logical places for IDPs to migrate to, part of what I have defined as Acehnese social capital. Additionally, there are aspects of Acehnese culture that made such patterns of mobility intuitive, both for those on the move, and those acting as "hosts". As discussed above, the movements of IDPs tapped into Acehnese understandings of mobility, travel and hospitality. The end result was that, during periods of acute crisis, people opted to move as a strategy for survival. Furthermore, this was clearly disrupted by the flow of relief and reconstruction aid following the tsunami. The geography of aid distribution had a direct influence on the movements of IDPs and created alternative locations to *gampongs*. In the following section, I look in more detail at two villages, to show the problems for Acehnese communities of operating outside of *gampong* infrastructure.

SPLIT COMMUNITY: A BREAKDOWN OF ACEHNESE SOCIAL RELATIONS

The Indonesian Government's decision to provide barracks or military-style temporary shelters for tsunami IDPs posed a major challenge to Acehnese social structures and cultural practices. While there were certainly questions about whether such barracks were culturally inappropriate and suitable for a range of reasons, I want to focus on basic issues of provision. I have found that a combination of limited space and the slow pace of building some

TABLE 7.2
**Distribution of Tsunami IDPs in "Conflict-prone" Areas in
Northeastern Aceh**

NO	Districts	Data Satkorlak/Dinsos*			Data from BPDE**/Dinsos		
		8-Jan-05	24-Feb-05	25-May-05	22-Jun-05	4-Jul-05	20-Jul-06
1	Bireuen	23,550	14,043	49,945	38,662	49,945	35,963
2	Pidie	55,099	32,067	82,612	48,456	81,532	79,188
3	Aceh Utara	28,470	28,113	33,004	32,761	30,511	31,667
4	Lhokseumawe	3,456	16,412		10,643	2,494	6,260
5	Banda Aceh	27,980	40,831	49,921	56,874	49,921	37,382
6	Aceh Besar	107,740	98,384	97,485	55,800	99,815	64,145
7	Aceh Timur	1,849	14,054	12,422	13,773	14,072	20,456
8	Langsa	10,227	2,806	6,156	1,052	6,156	1,401
	Total	258,371	246,710	331,545	258,021	334,446	276,462
	Percentage from Total IDPs in Aceh	71%	61.5%	64.5%	82.2%	58.4%	59.7%

*Satkorlak is Indonesia's chapter of disaster mitigation agency working during the emergency period following the tsunami in Aceh; **BPDE, Badan Pengelola Data Elektronik, local government electronic data clearing house.

of the barracks led to situations where villages and even families were split up. The division of IDPs from the same *gampong* into different barracks is troubling. There were also cases where some IDPs were moved into barracks while others from the same village were not, due to the limited number of barracks the government could provide. In both cases, the cohesiveness of local social structure (*gampong*) was put under pressure. In the following sections, I use two case studies, of the Al-Mukarramah neighbourhood of Punge Jurong and Lambung village (Gampong Lambung), to show how humanitarian interventions broke up the *gampong* concept and contributed to horizontal conflicts.

The Case of Al-Mukarramah Neighbourhood

Al-Mukarramah neighbourhood is one of five in the village of Punge Jurong in Banda Aceh. This neighbourhood was almost completely destroyed by the tsunami, even though it was situated nearly four kilometres inland. Prior to the tsunami, the neighbourhood had 3,812 members (382 households), of which only 875 survived. Up until June 2006, some 192 survivors lived in the Lhong Raya barracks. Around 165 survivors lived in barrack units and tents within the neighbourhood, while the rest were scattered in different host communities/families in Banda Aceh, Pidie and North Aceh, with some going as far as Medan, Jakarta and Malaysia.[33]

Days after the tsunami, survivors from this neighbourhood were scattered in several IDP camps in Banda Aceh. Some went back to Pidie to drop off their smaller children, women or the elderly in *gampongs* where they had family connections. When they got back to Banda Aceh in early to mid January to look for missing family members or to find humanitarian and relief supplies, they were largely based in a public building (*Gedung Sosial*). Led by Abubakar Ishak, a community organizer asked by the group to take the place of the missing neighbourhood head, the villagers seemed very united during the first months after the tsunami.[34]

However, when a temporary living centre (TLC) in the form of barracks was built by the government in April–May 2005, the number of units for IDPs from this neighbourhood were not enough. This led to an automatic division. There was immediate tension over who got to move to the barracks and who would stay in the public building. Fortunately, around that time, the government let IDPs return to their villages.[35] Mr Ishak led meetings with his people and decided that women and young children along with a few men would move to the barracks in Lhong Raya, about four kilometres from the neighbourhood and the public building. The rest decided to return to the destroyed but already accessible neighbourhood in Al-Mukarramah.

The groups that returned to Al-Mukarramah said that they were glad that they had not moved to the barracks, as the conditions in the barracks were terrible: poor and limited water and sanitation facilities and a lack of privacy were two common complaints I heard during field visits. While some people in the barracks admitted to the poor conditions, they in turn thought that people in Al-Mukarramah had to live with even worse conditions out on the open ground in tents and makeshift shelters, with no other infrastructure.

During the emergency period following the tsunami, there was a lot of confusion about what kinds of interventions humanitarian and relief organizations should make. This is in part because they were not certain whether the Indonesian Government would allow returnees to coastal areas and also because many of the barracks were not Sphere-compliant.[36] However, after organizations became more involved in providing support for IDPs, a number of problems emerged. This was especially the case with divided groups from the same *gampong*. Early returnees claimed that their fellow villagers living in barracks were well taken care of, as everything was provided by the government and relief organizations. In contrast, the people in barracks claimed that they did not get enough. Suspicion and distrust increased as some village figures competed for influence in the distribution of resources. During my fieldwork, it was made clear that arguments over the management of the relief and reconstruction process were amplified both by the disconnect between the separate "factions" of the *gampong* and because of the breakdown of the basic structures that people in the *gampong* had previously relied on for resolving disputes.

The culmination of the divide came when Pak Abu, the village head, resigned in early May 2006 at a neighbourhood meeting. He said that he was tired of being criticized by his fellow villagers living outside the neighbourhood (in barracks and host communities), who came back occasionally and had done nothing but demand supplies and services they already had in their temporary shelters. Furthermore, he said that despite his efforts to help return the neighbourhood to normalcy, some people suspected him of corruption, based on hearsay. He said that it was not easy to make everybody happy when they did not even meet regularly: some lived in the neighbourhood under tents, barracks and destroyed houses, some lived in the barracks outside the neighbourhood, some lived far away outside Banda Aceh and some lived in host communities in Banda Aceh.

This example highlights a number of significant issues that were common around tsunami-affected areas in the first year after the disaster. First, there was a breakdown in the central social mechanisms through which *gampongs* are governed, sometimes with the death of the *keucik*. Second, the members of the village were divided and physically separated. This had profound effects on their ability to make cohesive decisions that best represented the community as a whole. Finally, the provision of relief and resources by outside organizations was done outside of community frameworks and led to jealousy and tension amongst the village members. Much of this could have been avoided if a basic working knowledge of how *gampongs* operate had served as the framework for managing IDPs and aid distribution. I will now present the case study of Lambung, to demonstrate some of the factors that allowed that community to remain cohesive in spite of the damage from the tsunami.

Lambung: A Re-united Community

Gampong Lambung's success in participatory land consolidation has been applauded by the government of Banda Aceh and BRR. So much so in fact, that Gampong Lambung was chosen as a "model village" for reconstruction. Seventy billion rupiahs (about US$7.7 million) were earmarked by the BRR for developing forty-two blocks in the *gampong*, with a drainage system, telephone network and range of public and social facilities.[37] These projects were able to take shape in large part because members of the community were very flexible about the use of land in the village for public good, including voluntary efforts to share and consolidate land to support road-building.

Land certification and consolidation were particularly successful in this *gampong*, compared to Al-Mukarramah and other villages in Aceh. Both Lambung and Punge Jurong got support for land certification efforts from RALAS. The Reconstruction of Aceh Land Administration System (RALAS) supported the National Land Agency (BPN) funded by a US$28.5 million grant to "identify land ownership through a community-based adjudication process and issue land titles to up to 600,000 land owners".[38] As discussed by Fitzpatrick in this volume, the adjudication process started with community-mapping as an "important prerequisite for the reconstruction of settlements". By the end of 2006, this process was completed in Gampong Lambung, while the same process was not nearly as successful in Punge Jurong. While there was a problem of bureaucracy in the BPN and with the low capacities of implementing NGOs, whether a community could come together to solve its own problems was essential to the success of this programme.

More than 1,500 out of the nearly 1,900 residents of Lambung died in the tsunami. Lambung was almost totally "cleared" by the tsunami, leaving not a single intact structure, according to Zaidi M. Ada, *keucik* of Lambung.[39] As of 2006, there were 420 people registered as official residents in Lambung, of which about 400 were survivors from the quake and tsunami. The remaining were new residents brought in via marriage after the tsunami. Gampong Lambung has always been a *gampong*, and has never functioned as either a *desa* or *kelurahan*. This is the only village with a *gampong* structure in the subdistrict of Meuraxa.

Right after the quake and tsunami, survivors from Lambung took refuge in the hilly area of southeast Banda Aceh around the TVRI (state TV) tower and relay station in Mata Ie. This area became a well-known IDP camp, with persons from all over Banda Aceh and Aceh Besar making their way there. This was also the first major concentration of IDPs to become the focus of intensive government and non-government, local, national and international tsunami humanitarian and relief aid in the first few days after the tsunami. What started as a rather unplanned response to the tsunami began to formalize as an IDP "camp", due to the attentions of aid agencies.

Interestingly, survivors from Lambung did not stay in the TVRI camp. Instead, many took refuge with relatives who had several houses near the TVRI area. When more survivors made their way there, they got together and rented a piece of land in the area and erected several tents. This allowed them to stick together and also separated them from the bigger crowd in the TVRI camp. When asked why people from Lambung separated themselves from the TVRI camp, Hardiansyah, one of Lambung's community organizers, answered:

> We did not *separate* ourselves from others. But we *centralized* ourselves, so we can focus, reorganize, and mobilize our people easily. Everybody who was healthy and strong enough works together. First, we have these relatives offering us their places to stay, then when it was overcrowded, we rent land close to the place and erected several tents with public kitchen, water pump, and latrine". (my italics)[40]

They returned to their village as soon as improved road access allowed them to do so. They rejected being put into barrack units away from their village. Instead, they proposed that donors help early returnees like them by providing barracks on the sites of their destroyed villages. However, even before they received any outside assistance, they built one barrack unit by *gotong royong,* an Indonesian concept of self-help and community involvement. Survivors used salvaged materials and bought new material with donations from relatives in Banda Aceh and elsewhere. Later, an international organization built two more barracks in the village, allowing more survivors to return.

There were survivors from Lambung who did not stay together, but all agreed that there was only one "Posko"[41] for every survivor from Lambung to turn to when needed; unlike other villages whose survivors opened several Poskos, depending on where they took refuge. Survivors from Lambung spread into different host communities; some even temporarily left Banda Aceh. But when they called or returned, they had only one number to call and one place to return. All decisions on public affairs were made at the Posko, based on *mufakat.* The *gampong* officers were put back together as soon as they had first *rapat (mufakat) gampong* when they were still in the IDP camp. Therefore, survivors from Lambung re-organized faster and better than, for example, the survivors of Punge Jurong. This is well illustrated in the following observation by Hardi, the young community organizer from Lambung: "I noticed the weakness of other village is that they opened several 'poskos' to get the most from relief supplies, but caused problems of coordination".

Like many other communities, early returnees in Lambung were all male survivors. Female survivors stayed with host families and did not return until the two additional barracks were erected. Once temporary shelters in the village were available, however, women chose to leave the host communities or other types of temporary shelters and returned to the village where they found more acquaintances. When asked what made Lambung survivors re-organize faster, and maintain coherence, Hardi said:

> Sticking together with people of acquaintances who underwent the same experiences ease your burden and sadness. But leadership from our *keucik* and *sekdes* (*waki*) was also essential in keeping us together. Our *keucik* is an achiever (*pendobrak*) while our *waki* is a careful planner. But I think we have been very cohesive (*kompak*) since before the tsunami.

Sadly, however, Hardi observed that the initial cohesiveness began eroding after the emergency period, especially when "cash for work" programmes were introduced in Lambung by NGOs:

> Before many friends of my age and other fellow villagers were willing to work voluntarily, for free, for our own *gampong*. Now, there is still some willingness, but I felt that it has been gradually decreased. People started to be lazy to work together, *gotong royong*, maybe, since the introduction of 'cash for work'. It would be better if 'they' just give the money away, without attaching it to any scheme like that. No reason needed to 'work', cleaning up your own *gampong*. It should have been clear difference: donation or earned-money. Don't mix them up. Why they have to make up reasons to give away money?

CONCLUSIONS

Local reactions to the prolonged conflict attest to the ability of the Acehnese to cope with almost no outside intervention. Migration within and out of Aceh was one of the most important modes of survival for Acehnese during the conflict and continued to be immediately after the tsunami. Those population movements were supported by structures of social relations and networks that are deeply embedded within both *gampongs* and Acehnese understandings of mobility and hospitality. These in turn are rooted within both culture and Islamic teachings.

My research has made clear that displacement and return in Aceh was fluid, "instead of the static-ness of camp life as is often represented" in official reports. This is in agreement with Siapno, who found that the actual situation of IDPs needs to be discussed

> beyond the way this term displacement is often used in particular in human rights reports, i.e. enforced physical and geographic displacement, mostly in repressive and negative sense, to open up possibilities including fluidity and resilience, and ways of dealing courageously with forced displacement.[42]

Like Siapno, I have found that displacement was very much a mode of resilience and resistance for Acehnese, making it important to see them as more than just "the victims", "the oppressed" or merely uninformed "displaced masses". Additionally, my work has made it clear that the strategic movements of IDPs were generally based upon *gampong* connections and cultural understandings of mobility, and that this was challenged by the massive amounts of external intervention that followed the tsunami.

The humanitarian relief services centred in Banda Aceh and other urban areas after the tsunami caused indecisiveness among some IDPs about where to resettle during the first months following the catastrophe. Resettling in the rural interior regions where they had some connections and social relations, temporarily or otherwise, meant that IDPs could potentially miss out on relief aid provided by national and international organizations. This created a situation in which people had to leave both "host" villages where they sought temporary refuge and also the sites of their home villages that were destroyed by the tsunami, to be closer to services that were largely based in urban areas. Many IDPs worried that if they stayed too long in their rural villages, they might not get the supplies they needed, as described by Zulfikar (forty-six years old), a resident of Lingke in Banda Aceh:

> A couple days after the tsunami we all went back to our village in Pidie. Although we had to live at my mother-in-law's place with another twenty one persons from my wife's extended family who also lost their houses in Banda Aceh and Aceh Besar, it was good for us, especially the children, to get together with family and other villagers and feel safe with the support of one another. But after about a month, we realized that we could not get enough help if we stayed in Pidie. So all of us move back to Banda Aceh ... This way, we get relief supplies.[43]

As always, this raises the classic question about whether services should follow IDPs or vice versa. For the most part in Aceh, the latter was the case, in part because of very real logistical considerations, but also because relief agencies did not tap into *gampong* mechanisms when delivering aid. The presence of national and international humanitarian and then reconstruction organizations in Aceh created alternative temporary support systems for IDPs. Unfortunately, it seems at least outside of the main urban centres, this undermined key community structures.

The cases studies discussed make it clear that a number of factors contribute towards the successful maintenance of community. In particular, it seems that communities with strong leadership and social cohesiveness are better in navigating relations with outside intervention. In contrast, split communities fall behind in rehabilitation and reconstruction. I end by arguing that *gampongs* should have been the basis for reorganizing Acehnese communities after the conflict and tsunami. *Gampongs* have deep historical presences, attested social cohesiveness and democratic social structures. *Gampongs*, instead of subdistricts (*kecamatan*) or households, should have been the basis for rehabilitation and reconstruction and their coordination. Households are obviously too small and numerous to be the basis for coordination. *Kecamatans*, on the other hand, lack social virtues. These lessons can be applied outside of Aceh and Indonesia. It is important for aid and relief organizations to engage more fully with institutions that are firmly embedded within local communities. This allows external organizations to tap into pre-existing social capital, rather than disrupting it.

Notes

1. P. Curson, "Introduction", in *Population and Disaster*, edited by J.I. Clarke, P. Curson, S.L. Kayastha and P. Nag (Oxford: Basil Blackwell, 1989).
2. Even two years after the tsunami, Oxfam reported that about 70,000 people were still living in military-style temporary shelters called "barracks" in Aceh, while many more were still living with host communities throughout the region. <http://www.oxfam.org.uk/applications/blogs/pressoffice/2006/11/oxfam_calls_to_step_up_respons.html> (accessed 10 January 2007).
3. Clarke, Curson, Kayastha and Nag, *Population and Disaster*.
4. See, for example, P.M. Blaikie et al., *At Risk: Natural Hazards, People's Vulnerability, and Disasters* (New York: Routledge, 1994); V.T. Brook and B.P. Paul, "Public Response to a Tornado Disaster: The Case of Hoisington, Kansas", *Papers of the Applied Geography Conferences* 26 (2003): 343–51; T. Cannon, "Vulnerability Analysis and the Explanation of 'Natural' Disasters", in *Disasters, Development and Environment,* edited by A. Varley (Chichester: John Wiley & Sons, 1994), pp. 13–30; A. Lavell, "Opening a Policy Window: The Costa Rican Hospital Retrofit and Seismic Insurance Programs 1986–1992", *International Journal of Mass Emergencies and Disasters* 12, no. 1 (1994): 95–115; D. Parker et al., "Reducing Vulnerability following Flood Disasters: Issues and Practices", in *Reconstruction After Disaster: Issues and Practices,* edited by A. Awotona (Aldershot: Ashgate, 1997), pp. 23–44; K. Smith and R. Ward, *Floods: Physical Processes and Human Impacts* (New York, John Wiley & Sons, 1998).
5. B.K. Paul, "Evidence against Disaster-induced Migration: The 2004 Tornado in North-central Bangladesh", *Disaster* 29, no. 4 (2005): 370–85.
6. See Field, for example, for a "key ideas" summary of social capital. J. Field, *Social Capital* (New York, Routledge, 2003).
7. P. Bourdieu, and L. Wacquant, *An Invitation to Reflexive Sociology* (Chicago: University of Chicago Press, 1992).

8. R. Putnam, R. Leonardi and R.Y. Nanetti, *Making Democracy Work, Civic Tradition in Modern Italy* (Princeton, NJ: Princeton University Press, 1993); R. Putnam, "The Prosperous Community", *American Prospect* 13 (1993): 35–42; R. Putnam, "Bowling Alone: America's Declining Social Capital", *Journal of Democracy* 6, no. 1 (1995): 65–78; R. Putnam, *Bowling Alone: The Collapse and Revival of American Community* (New York, Simon and Schuster, 2000).

9. J. Siapno, "Living Displacement: Everyday Politics of Gender, Silence, and Resilience in Aceh", in *Conflict, Violence, and Displacement in Southeast Asia*, edited by E.L. Hedman (Ithaca, NY: Cornell University Press, 2007).

10. Snouck Hurgronje, *The Acehnese*, translated by A.W.S. O'Sullivan, Vols. 1 and 2 (Leiden: Brill, 1906), p. 58.

11. Some use *geusyik* or *geuchik*. Local Law (Perda) No. 7/2000 uses *geusyik*, Qanun No. 5/2003 uses *keuchik*.

12. S. Tripa, *Memahami Budaya dalam Konteks Aceh* [Understanding culture in the Aceh context], 16 May 2006 <http://www.acehinstitute.org> (accessed 10 December 2006).

13. Snouck, *The Acehnese*, 2: 314.

14. Siapno, "Living Displacement".

15. Ibid.

16. Hedman and Siapno are the only two scholars the author knows who have worked on this issue. Nah and Hyndman have reported on Acehnese diaspora in Malaysia and North America. J. Hyndman and James McLean, "Settling Like a State: Acehnese Refugees in Vancouver", *Journal of Refugee Studies* 19, no. 3 (2006): 345–60; Alice M. Nah, "Ripples of Hope: Acehnese Refugees in Post-Tsunami Malaysia, *Singapore Journal of Tropical Geography* 26, no. 2 (2005): 249.

17. "Towards a Mapping of 'At Risk' Ethnic, Religious and Political Groups in Indonesia*", Inside Indonesia* 1999 [cited 10 April 2006]; 86: <http://www.serve.com/inside/digest/ dig86.htm>.

18. G. Hugo, "Pengungsi — Indonesia's Internally Displaced Persons", *Asian and Pacific Migration Journal* 11, no. 3 (2002): 297–331.

19. Global IDP Project, *Internal Displacement: A Global Overview of Trends and Developments in 2003*, Geneva, February 2004. In this report, the entry is specifically written "Indonesia (Aceh)", which I assume is based on the IDPs situation in Aceh rather than a combined number throughout Indonesia. <http://www.internal-displacement.org/idmc/website/resources.nsf/(httpPublications)/21973CC905348E7F802570BB0042BB63?OpenDocument> (accessed 18 May 2009).

20. M. Ross, "Resources and Rebellion in Aceh, Indonesia", paper prepared for the Yale-World Bank project on "The Economics and Political Violence", 5 June 2003.

21 As reported in Priyambudi Sulistyanto, "Whither Aceh?" *Third World Quarterly* 22, no. 3 (2001): 437–52; p. 447.

22. www.reliefweb.int.
23. *Kompas*, 28 May 2003.
24. In E.-L.E. Hedman, "The Right to Return: IDPs in Aceh", 2006 <http://www.reliefweb.int/rw/RWB.NSF/db900SID/KHII-6PG5G6?OpenDocument> (accessed 18 May 2009).
25. See, for example, Human Rights Watch, "Malaysia: Stop Deportations of Acehnese Refugees", 1 January 2004 <http://www. hrw.org/english/docs/2004/04/01/malays8379.htm> (accessed 10 Aug 2006); Nah, "Ripples of Hope" and Hyndman and McLean, "Settling Like a State".
26. Based on the author's observations and several interviews in Penang and Kuala Lumpur in Malaysia and Medan in Indonesia in 2003 and 2004.
27. See Appendix 1 for a copy of the letter.
28. There are three main resources for data on IDPs in Aceh after the tsunami: (1) Satkorlak, the provincial government agency for disaster management which produced IDPs estimates in the first weeks after tsunami. Satkorlak, in collaboration with Aceh's Dinsos (Social Affairs Office) derived the numbers "from a combination of general estimates of population figures and data collected from the sub-district heads (*Camats*)"; (2) Aceh's Electronic Data Management Agency (BPDE) with its field surveyors from Garansi, an Indonesian NGO: "Garansi was contracted to register IDPs throughout the province and adopted an approach of finger printing IDPs to prevent double counting"; (3) Population Census for Aceh and Nias (Sensus Penduduk Aceh dan Nias, SPAN) by Office for Statistics (BPS) Aceh, which launched its partial results on 29 November 2005. The foreign donor-funded SPAN reported somewhat lower figures of IDPs compared to the first two. But Cibulskis concluded that "there is some comparability between SPAN and other sources of data, though not perfect partly because of different definitions and timing".
29. Satkorlak and Dinsos. SPAN reported 192,055 IDPs during August–September 2005 while BPDE/Dinsos reported 436,820 IDPs, as of 8 September 2005.
30. Siapno, "Living Displacement".
31. When the SPAN is completed, the number of "ex-IDPs" is somehow lower, as seen in Table 1.
32. A complete table of tsunami IPDs distribution in all Aceh districts is in Appendix 2.
33. Author interview with Abubakar Ishak, head of Al-Mukarramah neighbourhood, Punge Jurong, February 2006.
34. The author visited the shelter several times in January–April 2005.
35. In the immediate aftermath of the tsunami, the government created buffer zones along the coastal regions that were hardest hit by the tsunami. This was done in part because of the massive damage done, fear of aftershocks and additional tsunamis, and to consolidate people who needed emergency humanitarian aid.
36. Guiding Principles on Internal Displacement.

37. "Untuk Bangun Desa Lambung Butuh Dana Rp 70 M", *Serambi Indonesia Daily*, 15 October 2006.
38. Information on this project is available on the World Bank website at the following link: <http://web.worldbank.org/WBSITE/EXTERNAL/ COUNTRIES/EASTASIAPACIFICEXT/INDONESIAEXTN/0,, contentMDK:20877372~pagePK:141137~piPK:141127~theSitePK:226309,00. html> (accessed 30 March 2007).
39. Government of Lambung presentation at Aceh Habitat Club, a reconstruction participatory forum hosted by UN-Habitat and The Aceh Institute.
40. Interview with Hardiansyah, 29 January 2007.
41. "Posko" is short for *Pos Komando*, which existed during the early emergency period in Aceh to act as mediators between villagers and aid agencies.
42. Siapno, "Living Displacement".
43. Interview with Zulfikar (not a real name) in Banda Aceh, 10 March 2006.

APPENDIX 7.1
Letter from United Nations Recovery Coordinator for Aceh and Nias (UNORC) to the Head of Aceh and Nias Rehabilitation and Reconstruction (BRR) on IDP definition

United Nations   Nations Unies

OFFICE OF THE RECOVERY COORDINATOR FOR ACEH AND NIAS

According to the SPAN, the total number of current IDPs amounted to 192,055 in Aceh during August and September 2005 when fieldwork was undertaken. It is difficult, however, to arrive at consistent figures on IDPs for Nias as the tables presented in the slide presentation of the census varied from other hand outs provided during the SPAN presentation by the Bureau of Statistics (BPS) on 29 November 2005.

The figures for Aceh in any case are far lower than the numbers previously produced by Satkorlak and Sinsos. This seeming discrepancy may be attributed among other things to differences in definition addressed below.

The definition of IDPs in the SPAN census is based on whether respondents at that time identified themselves as an IDP. The question posed to respondents was "*Apakah saat ini anda sebagai pengungsi?*" Engl: Are you an internally displaced person at this time? This question was necessarily subjective, as data were collected through questioning household heads on whether they self identified as IDPs.

Significantly, the census also records that 312,463 tsunamis/earthquake survivors who once identified themselves as IDPs no longer do. The follow up question that was posed to respondents who answered no to the previous question of whether they were an IDP was "*Apakah Anda pernah mengungsi setelah gempa/tsunami?*" Engl: Have you ever been displaced as a result of the earthquake/tsunami?

This large number of Tsunami-earthquake survivors having once but no longer identifying themselves as current IDPs does not imply that they automatically no longer require support from relief and recovery agencies in Aceh and Nias, quite the contrary. For example, many survivors are not recorded as IDPs, but are still living in tents or ad hoc structures on their own lands.

There may also be cultural, psychological or linguistic reasons why such a large number of people who once considered themselves IDPs no longer want to identify themselves as such. Two possible explanations stand out. Firstly, many tsunami-earthquake survivors have gone back to their own lands and now consider themselves as having returned home and therefore no longer an IDP regardless of the living conditions they experience. Secondly, being labelled a "pengungsi" or IDP in Aceh may carry a negative connotation which many tsunami survivors probably want to avoid.

While we maintain the primacy of census data over other sources, these issues suggest a need to review our usage of the category of IDPs in the Aceh-Nias case. In fact, focusing on IDP figures can be quite misleading, particularly if it is interpreted to mean that they are the only ones who need relief and other types of support.

It is evident from the 312,463 people in Aceh who define themselves as former IDPs that few of them have obtained proper shelter in the form of new houses nor have many had their livelihoods fully restored. Simply put, the distinction between current IDPs and former IDPs is not meaningful in terms of planning to address demands for assistance.

All of the above calls for a review on the kind of data categories that should be used to aid effective strategic planning and decision making for the recovery of Aceh and Nias. It argues perhaps for abandoning the label of internally displaced persons (IDPs) altogether, which is seriously limiting the classifications through which we define affected communities in Aceh and Nias. An alternative to this category is crucial to ensure that gaps in support to vulnerable communities are identified, uneven patterns of aid distribution are avoided, the potential for conflict reduced, and an integrated process of reconstruction in Aceh and Nias is promoted.

In the longer term, the recovery process will require the use of data categories that cut across many of the boundaries that have been taken as given between IDP and non-IDP groups and between tsunami affected and non-affected populations. Revised data categories should support a focus on sustainable livelihoods, poverty alleviation and longer-term development of Aceh and Nias. This will also enable better linkages between Tsunami programs and post-conflict reconstruction.

The launching of the UNFPA-supported census is the first step in collecting and analyzing adequate data to support the emerging needs of recovery and reconstruction.

Further analysis based upon cross-tabulation of the census data is currently being undertaken by BPS and partner agencies. This step should provide new analysis and new categories on the tsunami survivors, vulnerable population groups and potential developing beneficiaries while identifying key variables on needs and present living conditions.

Yours Sincerely,

Eric Morris
United Nations Recovery Coordinator for Aceh & Nias

APPENDIX 7.2

Distribution of Tsunami IDPs in all 21 Districts in Aceh

NO	Districts	Data Satkorlak/Social Affairs Office			Data from BPDE		
		8-Jan-05	24-Feb-05	25-May-05	22-Jun-05	4-Jul-05	20-Jul-06
1	Bireuen	23,550	14,043	49,945	38,662	49,945	35,963
2	Pidie	55,099	32,067	82,612	48,456	81,532	79,188
3	Aceh Utara	28,470	28,113	33,004	32,761	30,511	31,667
4	Lhokseumawe	3,456	16,412		10,643	2,494	6,260
5	Banda Aceh	27,980	40,831	49,921	56,874	49,921	37,382
6	Aceh Besar	107,740	98,384	97,485	55,800	99,815	64,145
7	Aceh Timur	1,849	14,054	12,422	13,773	14,072	20,456
8	Langsa	10,227	2,806	6,156	1,052	6,156	1,401
	Subtotal Northeastern Aceh	258,371	246,710	331,545	258,021	334,446	276,462
9	Aceh Jaya	31,465	31,564	40,422	20,781	40,422	33,309
10	Aceh Selatan	0	16,049	16,148	10,370	16,149	14,421
11	Aceh Barat Daya	3,180	13,847	3,480	4,599	3,480	4,062
12	Aceh Tamiang	0	800	3,396	1,261	3,396	1,792
13	Aceh Barat	56,497	49,310	72,689		72,689	68,931
14	Nagan Raya	10,712	11,281	17,040	12,679	17,040	13,514
15	Aceh Tengah	3,454	5,161	5,288	1,642	5,288	832
16	Bener Meriah	0	1,204	819		819	110
17	Aceh Tenggara	0	1,759	809	169	809	484
18	Aceh Singkil	0	2,032	105		30,967	21,884
19	Gayo Lues	0		158		158	4
20	Simeulue	0	15,551	18,009		42,751	24,400
21	Sabang	0	5,633	3,712	4,263	3,712	2,685
	Subtotal Central & Southwestern Aceh	105,308	154,191	182,075	55,764	237,680	186,428
	Total	363,679	400,901	513,620	313,785	572,126	462,890

8

ACEH'S FORESTS AS AN ASSET FOR RECONSTRUCTION?

Rodolphe De Koninck, Stéphane Bernard and Marc Girard[1]

AN EXCEPTIONALLY RICH FOREST HERITAGE

Recollections from the Seventeenth Century

Forests and forest products have played a central role in Aceh's history; Aceh's very name, according to Denys Lombard (referring to William Marsden), may even be attributable to a plant.[2] During the first decades of the seventeenth century, when Aceh was ruled by Sultan Iskandar Muda and had reached its apex as a regional power, its forest resources played an important economic role. According to Lombard, who quotes the *Hikayat Atjéh*, the kingdom's prosperity was at least partially linked to its control over resources of the forested interior, among them jungle animals and numerous plants and wood products.[3] Exports of these products continued well after the reign of Iskandar Muda.[4] They included elephants — those same elephants that had played such a crucial role in insuring Aceh's military and even naval power[5] — as well as horses and an apparently even greater number of forest products than before.

Ever since those glorious days in Acehnese history, as settlements and agriculture gradually occupied much of the coastal lowlands, with some expansion even into the interior, Aceh has remained predominantly forested, more so than the rest of Sumatra, at least until very recently.

The Leuser Ecosystem

Mainland Aceh covers some 55,400 km^2 (Figure 8.1). The core of that territory,[6] nearly 80 per cent of it, stands over 100 metres above sea level and is characterized by "splayed out ranges and valleys".[7] That mountainous terrain is predominantly the domain of the forest, a forest that constitutes a treasure trove of biodiversity. In fact, well over half of Nanggroe Aceh Darussalam is occupied by two exceptionally rich and largely mountainous forest domains: the Ulu Masen and Leuser ecosystems (Figure 8.2). The former, which extends over some 7,000 km^2, lies entirely within Aceh. Covering nearly 26,000 km^2, the much larger Leuser Ecosystem overlaps into the province of North Sumatra, which holds approximately one-eighth of it.[8]

The Leuser Ecosystem, with its impressive biodiversity, is the better known of the two. The area covered by the Leuser Ecosystem is said to be inhabited by more than 120 species of mammals, including the Sumatran elephant, Sumatran tiger, the endangered Sumatran rhinoceros and the Sumatran orang-utan, along with some 8,500 plant species.[9] Two of the more famous ones are the *Rafflesia*, the world's largest flower, and the *Amorphophallus*, the world's tallest. Equally important is the splendour of the mountainous landscape — including several peaks over 3,000 metres high, the loftiest being Mount Leuser, reaching over 3,400 metres — which characterizes much of the area, with deep valleys surrounded by steep, tree-covered slopes and volcanic forms, along with numerous streams and waterfalls. In fact, the ecosystem acts as an immense water reserve that feeds most of the coastal lowlands, the vast majority of Aceh.

The importance of the environmental and biological heritage represented by the Leuser Ecosystem was officially recognized in the early 1930s by the Dutch colonial administration, when the Gunung Leuser Wildlife Reserve was established. In 1980, the reserve was extended and given the name Gunung Leuser National Park (GLNP). Covering nearly 9,000 km^2, it nevertheless occupies only a fraction of the much larger Leuser Ecosystem.[10] The GLNP is, of course, largely mountainous, with nearly nine-tenths of its surface area lying above 600 metres.[11] Together with Kerinci Seblat and Bukit Barisan Selatan, the two other Indonesian National Parks also located on the Barisan Mountain Range of Sumatra, it makes up the Tropical Forest Heritage of Sumatra (TRSH). The TRSH is known for the exceptional biodiversity of its forests, a high level of endemism and even for possessing "significantly higher mammal diversity than the island of Borneo", as well as incorporating an important volcanic component.[12]

FIGURE 8.1
Aceh. Districts and Major Towns 2005

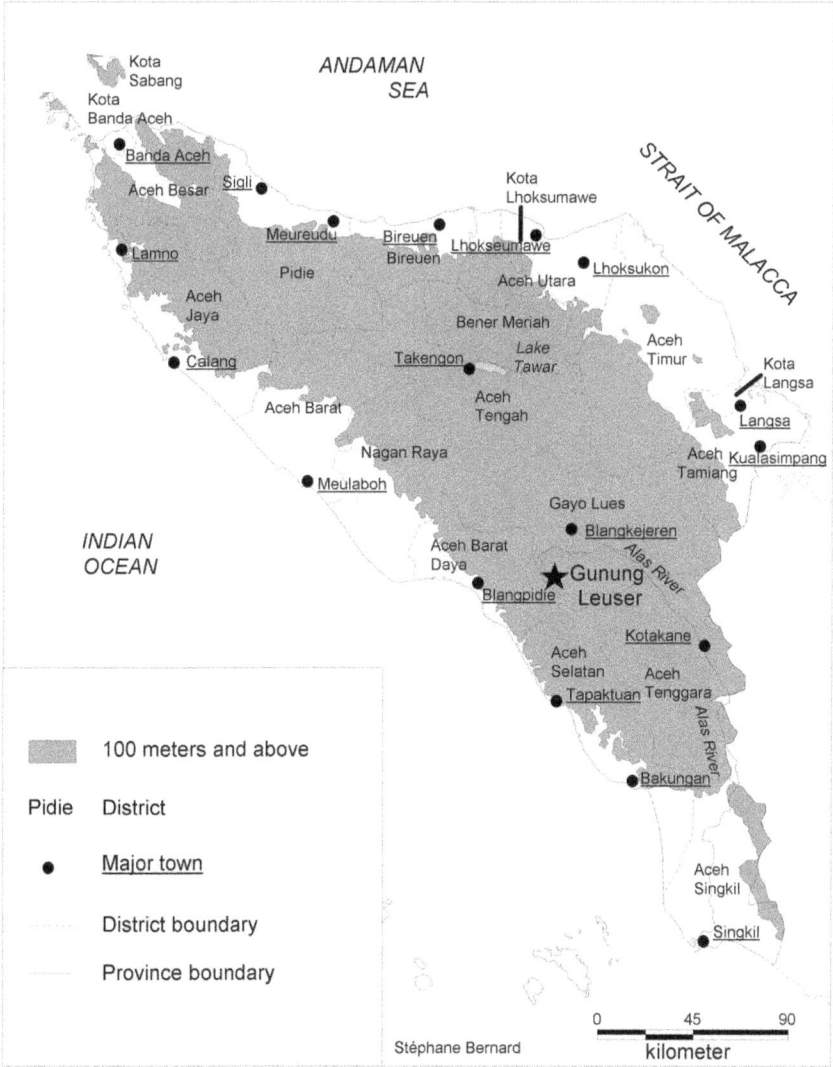

Sources: The Digital Chart of the World; Office for the Coordination of Humanitarian Affairs (OCHA) and Humanitarian Information Centre; De Koninck et al. (1977).

FIGURE 8.2
Aceh. The Ulu Masen and Leuser Ecosystems 2005

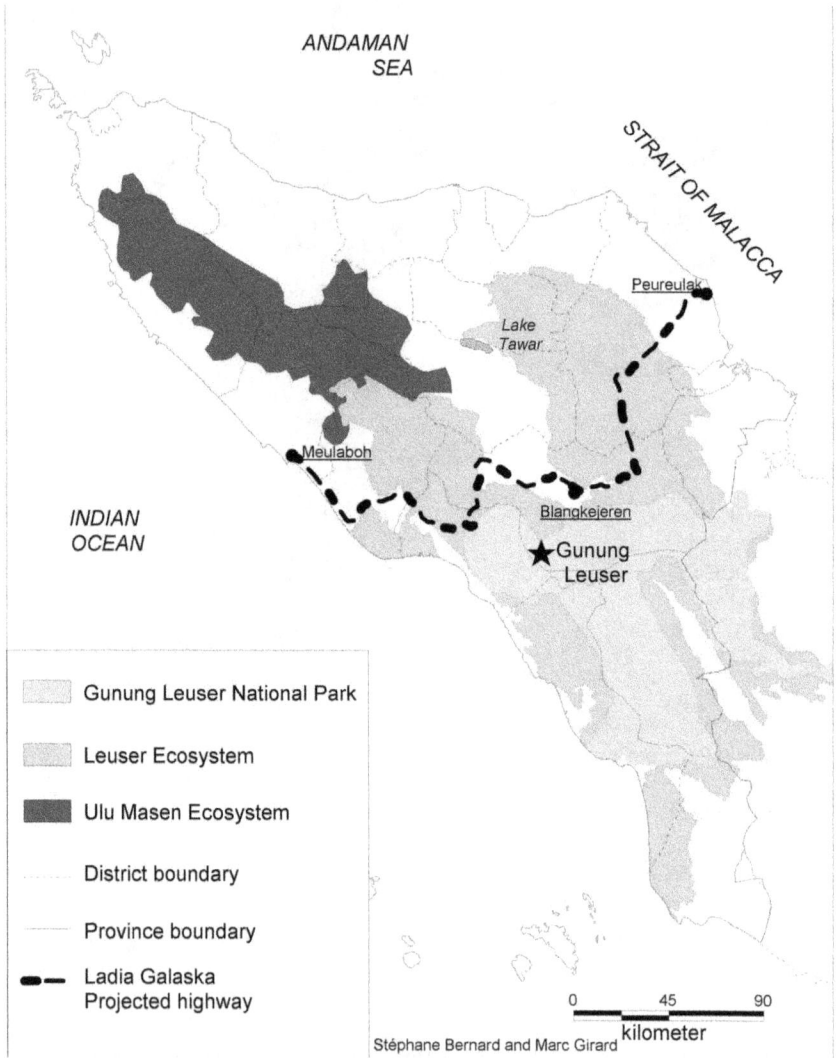

Sources: The Digital Chart of the World; The Leuser International Foundation; Office for the Coordination of Humanitarian Affairs (OCHA) and Humanitarian Information Centre.

The Leuser Ecosystem along with the GNLP were given further recognition in 1994 when the Leuser International Foundation (LIF) was established. In 1998, the LIF was given, by presidential decree, a thirty-year mandate to "manage the Leuser Ecosystem". By then, it was already apparent that the forest heritage of Aceh,[13] within and without the Leuser Ecosystem, was seriously threatened through increasing deforestation.

THE LAST SUMATRAN FRONTIER?

A Secluded Sumatran Province

At least since the 1960s, excessive deforestation has been a serious problem throughout Southeast Asia, particularly in Indonesia, where the country's forest heritage has been threatened by the combined effects of agricultural expansion and massive logging, legal as well as illegal.[14] The island of Sumatra, whose southern provinces had been for decades the prime destination for Javanese transmigrants, appeared singularly affected. Various diachronic and schematic representations of forest cover for the whole of Sumatra illustrate quite clearly the intensity of the retreat of the forest as well as Aceh's relatively privileged situation on the margins of the devastation process.

In a recent publication made available on the web,[15] the Sumatran Orangutan Society reproduced a World Wildlife Fund image which included a series of very sketchy yet quite revealing maps illustrating the evolution of Sumatra's forest cover between 1940 and 1996 (Figure 8.3).

Although, the source is very laconic and does not specify the nature of the forest cover represented, it most likely corresponds to primary forest. The resulting message is definitely clear: the forest domain of Sumatra has shrunk tremendously during that half-century interval, a fact amply corroborated by other sources.[16] In 1940, only a few regions appeared largely denuded of their original forest cover: these included the plantation belt around Medan, in North Sumatra province, and the heart of Southern Sumatra, around Palembang. By 1980, the retreat of the forest had accelerated noticeably, particularly in the same provinces, with Aceh still appearing relatively untouched, except along its eastern coastal lowlands.

However, the 1990 map points to a dramatic acceleration of the deforestation process, with nearly all regions and provinces being markedly affected. Although Aceh and Bengkulu still appear less devastated, the onslaught has definitely reached them. In the case of the more secluded Aceh, even if its forest cover still appears much more extensive, in relative terms, than that of all other provinces, its retreat has obviously begun in

FIGURE 8.3
Sumatra. Forest Cover Evolution 1940–96

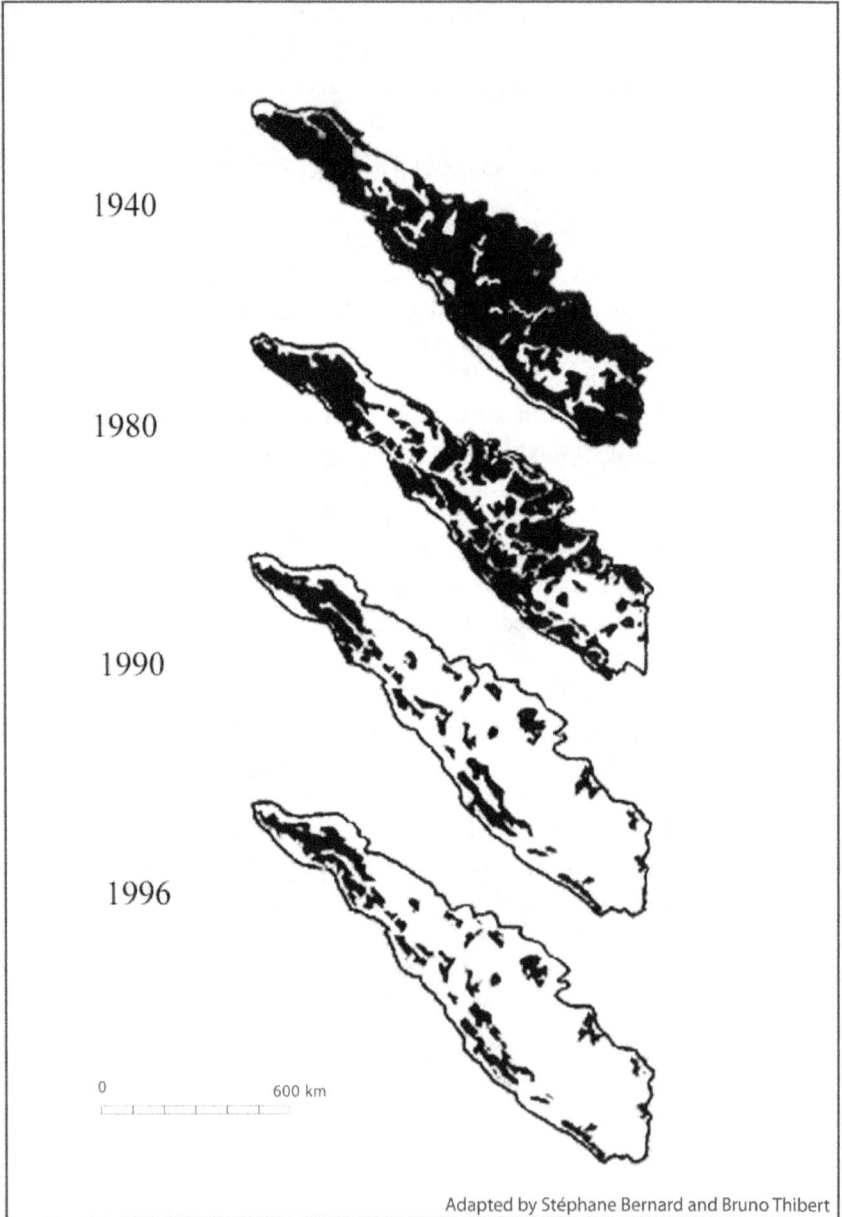

1940

1980

1990

1996

0   600 km

Adapted by Stéphane Bernard and Bruno Thibert

Source: www.orangutans-sos.org

earnest. This is confirmed by the 1996 map which illustrates strikingly the extent of contemporary pan-Sumatra deforestation. The noticeable difference between Aceh and the rest of Sumatra, regarding the extent of deforestation, is confirmed by figures provided in "The State of the Forest: Indonesia", published by Forest Watch Indonesia and Global Forest Watch.[17] By 1997, while the overall forest of Sumatra was down to 35 per cent of total land area, in Aceh it still stood at an apparently healthy 63.6 per cent.[18] In the southernmost provinces of the large island, the ratio was a very low 10.6 per cent in Lampung and 12.2 per cent in Southern Sumatra.

The acceleration of the deforestation processes in the 1980s is corroborated by other sources, first and foremost T. C. Whitmore's remarkable "Introduction to Tropical Rain Forests". In this publication, two maps of Sumatra's primary forest cover are juxtaposed:[19] these refer to the situation in 1980 and in the mid-1980s (Figure 8.4).[20]

The 1980 map is exactly identical to the one used by the Sumatran Orangutan Society and reproduced above in Figure 8.3. The pace of the retreat of the Sumatran primary forest cover, illustrated by the juxtaposition of these two maps, seems almost unbelievable. Even in Aceh, the disappearance of large chunks of forests is clearly shown, particularly in the heart of the mountainous interior.

The differential retreat of forest cover throughout Sumatra is corroborated by yet another form of diachronic mapping, this time applied to the entire Southeast Asian realm. In 1996, Bernard and De Koninck published two maps of overall land use in Southeast Asia. Concerning the years circa 1970 and circa 1990, these represented a synthesis of maps drawn from several sources.[21] Here we limit ourselves to the reproduction of the Sumatran sections of those two maps, with a single land use category being represented, that of forests (Figure 8.5).

This land use category includes all forest types, whether primary or secondary. The resulting illustrations are somewhat closer to the reality on the ground, to the extent that they allow for a somewhat more generous representation of forest cover. Nevertheless, the maps corroborate the ones examined previously (Figures 8.2 and 8.3). This is particularly true with Aceh, where the breaking up of the continuous forest cover in the interior as well as the retreat of the western coastal forests is clearly illustrated.

Finally, for heuristic purposes, we reproduce here a map representing the state of land use in Aceh circa 1966 (Figure 8.6).

This map was put together in the early 1970s by De Koninck, who relied on a number of manuscript maps then available only in Banda Aceh. It was later published in a research report.[22] A key feature illustrated by this map

FIGURE 8.4
Sumatra. Forest Cover Evolution 1980–c.1985

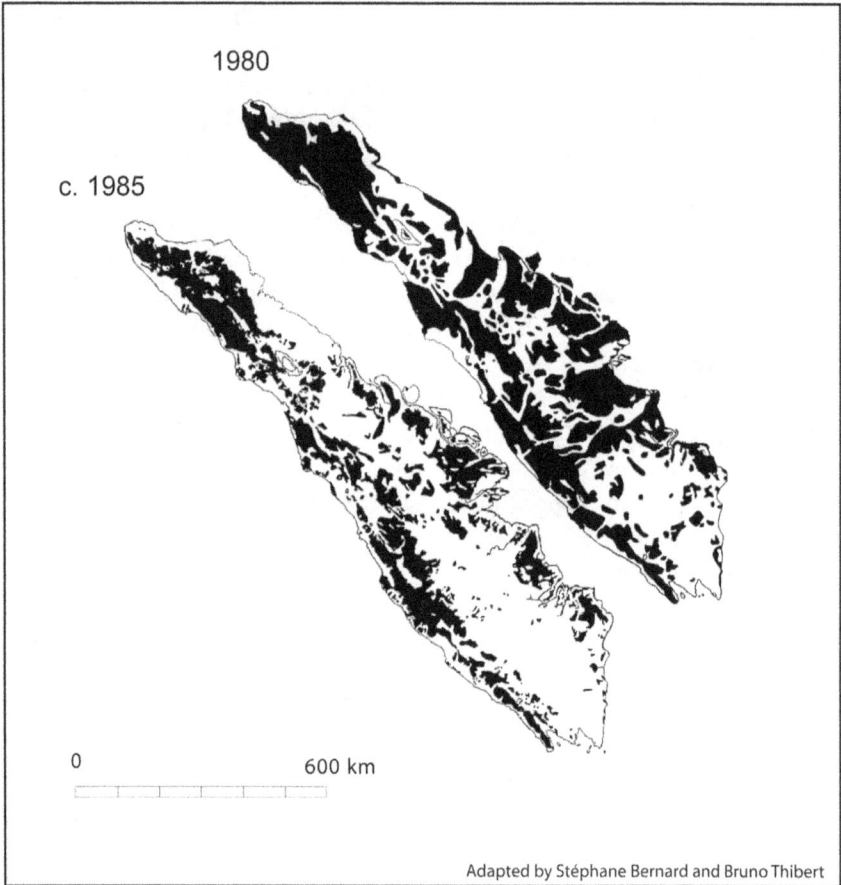

1980

c. 1985

0   600 km

Adapted by Stéphane Bernard and Bruno Thibert

Source: Whitmore, 1984 and Laumonier et al. 1983, 1986

of the dominant forms of land use is the overall importance of the Acehnese forest cover in the mid-1960s. Then, all forest types covered slightly more than eighty per cent of mainland Aceh.[23] Coastal lowlands appeared predominantly devoted to paddy cultivation, with noticeable expanses of commercial crop cultivation, such as rubber in Aceh Timur and coconut on the west coast, with coffee being confined to the highlands of Aceh Tengah.

FIGURE 8.5
Sumatra. Forest Cover Evolution c.1970–c.1990

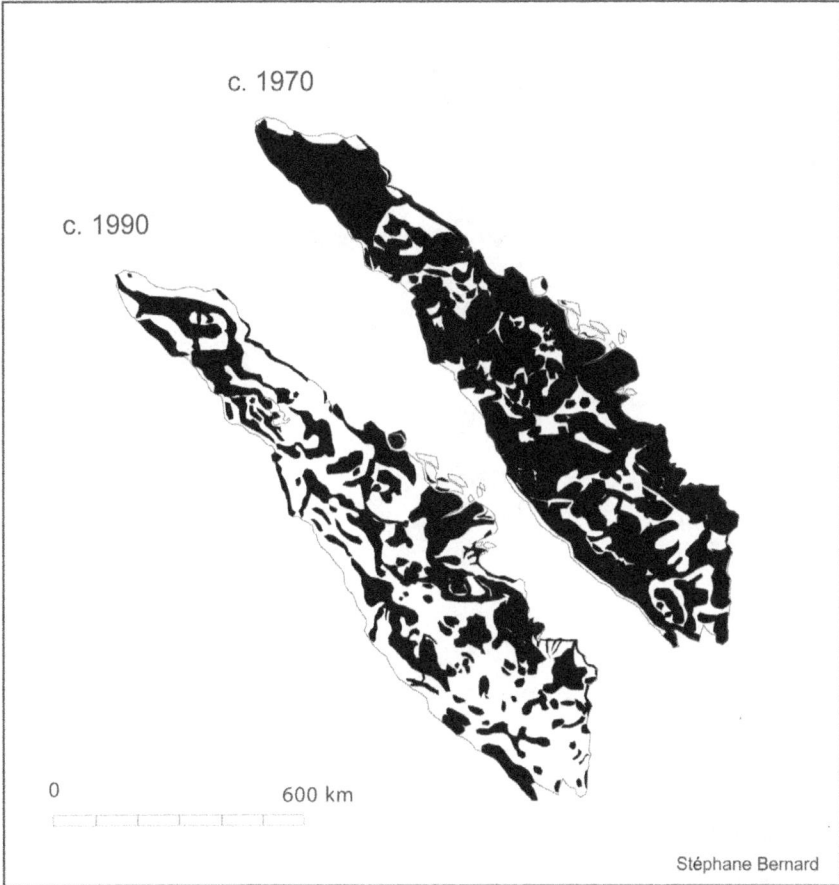

Sources: Collins 1980; Collins et al. 1991; World Atlas of Agriculture 1969; Weltforstat Atlas 1971; Whitmore 1984.

A Recent Conversion of Onslaughts

Until the 1960s and even the early 1970s, for a number of historical and geopolitical reasons, Aceh remained largely marginal within Indonesia. Its economy was still overwhelmingly based on subsistence agriculture, its modest export sector largely confined to coffee, rubber and very limited amounts of palm oil. Its population, which stood at two million in 1971, was primarily

FIGURE 8.6
Aceh. Land Use c.1966

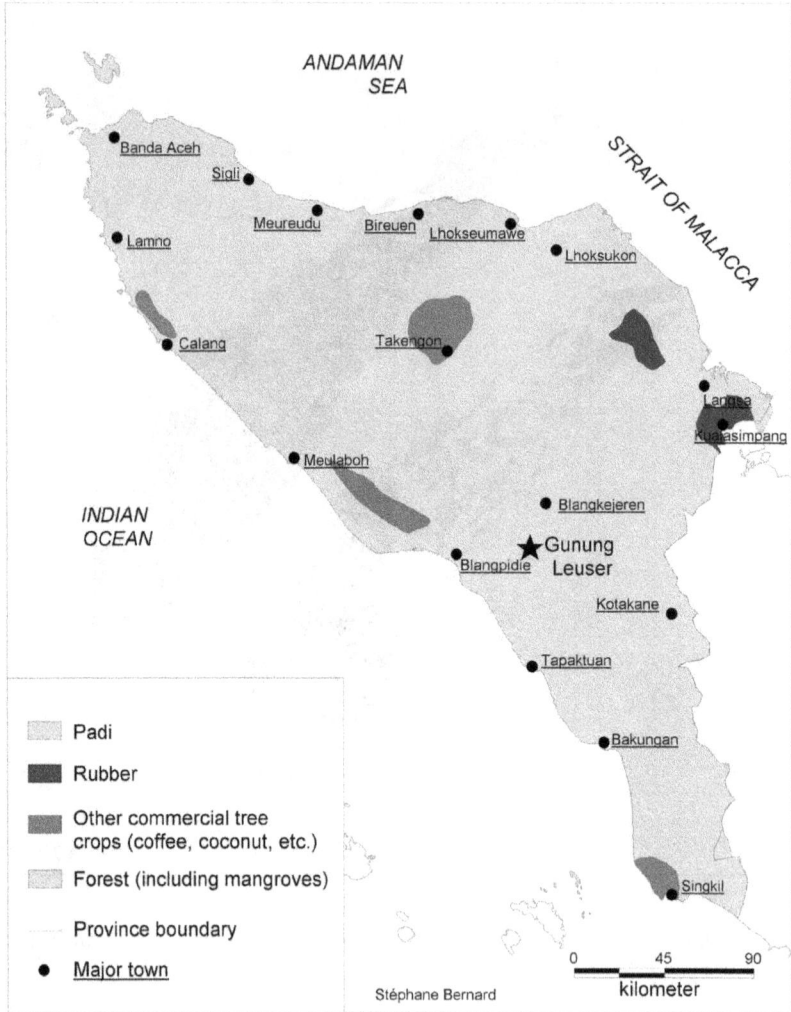

Source: De Koninck, Rodolphe, Gibbons, D.S. and Hasan, Ibrahim (1977) The Green Revolution, Methods and techniques of Assessment. A Handbook of a Study in Regions of Malaysia and Indonesia.

distributed along the coastal plains, especially the eastern ones, with two noticeable exceptions. In the uplands, two relatively important population clusters were located. The first centred on the town of Takengon, in the Lake

Tawar basin of Aceh Tengah district; the second was located in the town of Kotacane and along the Alas River, in the district of Aceh Tenggara, near the border with North Sumatra province.

Despite the fact that this general population distribution pattern has not altered significantly since then, even though total population has officially nearly doubled between the 1971 and 2000 censuses, the economic and political scenes have. The 1970s witnessed the beginning of the exploitation of the eastern coast's very rich gas fields[24] and the launch of the GAM insurgency.[25]

In other words, the "special province" became the object of increased economic and political attention, which was to greatly influence the destiny of the land. Although not everything that has occurred in Aceh since then — including the intensification of the deforestation process — can be attributed solely to these two series of events, they did signal the beginning of a new era, including for the forest heritage.

Due to the insurgency, the Indonesian military presence was increased. Even if, between 1976 and 2005, "the conflict waxed and waned in level of intensity",[26] the Indonesian army's business interests in the province never waned, particularly in logging, mostly in its illegal form. However, revenues from illegal logging did not only accrue to members of the army (Tentara Nasional Indonesia, or TNI), but also to GAM insurgents. More importantly, in a manner reminiscent of a "formula" applied extensively throughout the archipelago, huge forest concessions were granted to friends of the Suharto regime, including in supposedly protected areas. This was the case with substantial portions of the forested interior, including within the Leuser Ecosysem boundaries.

Not only did logging, both legal and illegal, then become the object of noticeable acceleration, but so too did agricultural expansion, here again in a manner reminiscent of a dual process well known throughout Indonesia, if not most of Southeast Asia.[27] Logged forests are taken over for agricultural production, whether by individual migrants or by large plantations. But in Aceh, while transmigration did occur, it never took on the importance it reached in other regions of Sumatra, particularly in the southernmost provinces. However, oil palm plantations did, as well as industrial tree plantations, the latter essentially for pulpwood. Oil palm cultivation, only modestly present in Aceh until the late 1980s, has obviously expanded tremendously since then. Although Aceh covers less than three per cent of Indonesia's territory, by 1999 the province was already "producing about 400,000 tons of palm oil, corresponding to some seven per cent of the national total".[28] As for the development of industrial tree plantations, it has been accompanied by the construction of paper mills, mostly owned by non-Acehnese. In addition, the rapid expansion of large-scale agricultural and industrial activities in ecologically fragile mountainous regions brought about a succession of environmental disasters, with floods and landslides being reported more and more frequently in several districts.

This being said, between 1996 and 2001, the expansion phase went through a respite, with an actual reduction in the size of the area devoted to plantations, particularly industrial tree plantations, at least within the confines of the Leuser Ecosystem (Figure 8.7). Nevertheless, encroachment does not seem to have been entirely interrupted, with new oil palm cultivation being reported even within the limits of the Gunung Leuser National Park.[29]

Overall, it seems quite evident that the opening up of the Acehnese economy to what Saiful Mahdi calls "mega projects" did result largely from the development of the very lucrative oil and gas industry; lucrative at least

FIGURE 8.7
Aceh. Encroachment on the Leuser Ecosystem 1996–2001

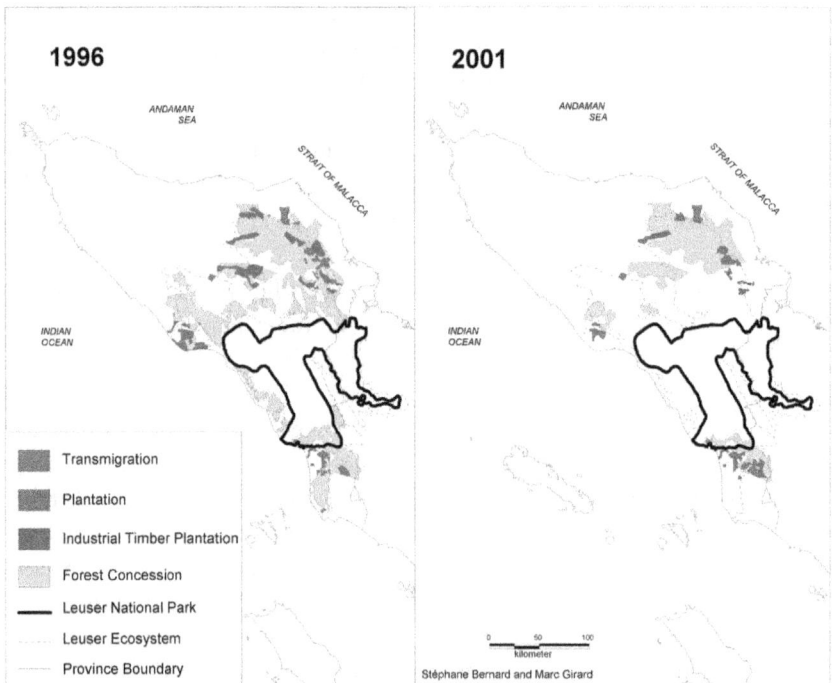

Sources: The Digital Chart of the World; Leuser International Foundation.

for the central government and for private contractors but not so much for the average Acehnese, given the relatively small number of local jobs provided by the industry once it was up and running.

THE CASE OF THE LADIA GALASKA HIGHWAY PROJECT

Among these mega ventures, the Ladia Galaska highway project stands out.[30] Debate about this road project, meant to link the west and east coasts through the mountainous interior, has raged ever since it was initiated by Ibrahim Hasan, governor of Aceh from 1985 to 1993. Actually, work on the 1,650 kilometre-long road network began only during the first years of the governorship of Syamsudin Mahmud (1993–2000). The so-called highway is made up of sixteen sections, some of which, particularly in the lowlands, were already quite passable before work began, while others were passable only during the drier seasons and then only to four-wheel-drive vehicles. The project consisted of linking together and, more importantly, upgrading a number of these pre-existing roads and tracks. By 1998, some 1,180 km or over seventy per cent of the highway was said to be complete, although much of that was not yet asphalted. That year, construction was interrupted, but apparently it has since resumed.[31]

The main problem with the Ladia Galaska highway, of which maps are hard to find,[32] is that it cuts through the heart of the Leuser Ecosystem. Its opponents point to the fact that it will fragment the ecosystem. They claim, with good reason, that it will necessarily facilitate illegal logging activities as well as poaching, thus endangering not only the forest and its biodiversity but the very services they render, particularly water supply for the lowlands. Also at stake are two indigenous groups, the Gayo and Alas, whose homeland is traversed by the highway.

A SERIOUSLY ENDANGERED HERITAGE?

Besides the maps we have assembled illustrating the exceptional acceleration in the rate of deforestation in Aceh, many statistics can be found pointing to the same conclusion. According to Eye on Aceh, forests still covered 68.5 per cent of the province's total area in 1989. By 1997, this proportion had been reduced to 63.7 per cent and, by 2000, to 48.5 per cent. Over that three-year period (1997–2000), the province lost nearly 860,000 hectares of forest.[33] This would mean that, over the eleven-year period (1989–2000), the average annual rate of deforestation reached nearly 7.0 per cent! That seems phenomenal, even by the standards of Indonesia, a country that, over the last four decades, has set all sorts of records regarding rates of environmental devastation. Whether or not these figures are realistic,[34] they do corroborate to a large extent the results that we have obtained, through GIS methods applied to the maps we have assembled (see Figures 8.3, 8.4 and 8.5).

In short, while by the late 1960s and early 1970s Aceh could still be considered as the last Sumatran forest frontier, that status has since definitely been lost. The rich forest heritage of the "special province", so tied to its cultural specificity and its relative geographical remoteness, now appears very seriously threatened.

CURRENT DRAWBACKS AND CHALLENGES

This brings us to the real question: is it threatened to the point of no return, to the extent that forest resources can hardly be harnessed, both for the reconstruction of the province after the devastation wrought by the December 2004 tsunami, and for further development?

Testimonies abound as to the very high demand for timber for the sake of reconstruction.[35] In May 2006, new logging permits were issued for the first time since 2001. However, not only has the harvesting of local timber now reached unsustainable levels, it appears largely insufficient to answer the demand. The oil palm expansion, as in several other regions of Indonesia, seems to go unabated. And there is ample evidence that illegal logging is still rampant, access to the more remote forested areas having been facilitated by the cessation of hostilities, which followed the July 2005 ceasefire brokered between GAM and the TNI. In other words, if the thirty-year insurgency did limit the possibilities of illegal logging — and even that is far from certain — all obstacles have now apparently been lifted. Consequently, renewed deforestation has already been identified as the major factor behind the spate of deadly floods and landslides that occurred in late December 2006 in at least five districts — Bireuen, Aceh Utara, Aceh Temiang, Gayo Lues and Bener Meriah — and said to be the worst in the last ten years.[36]

The interpretation of recent Google Earth images of Aceh seems to confirm the continuation of the retreat of the forest, including within the Leuser Ecosysem and GNLP, and with particular intensity in the Gayo Lues district, heartland of the Gayo minority (Figures 8.8 and 8.9).[37]

Of course, this elementary and crude form of monitoring Aceh's land use cannot suffice. As the numerous agents and agencies involved in Aceh's reconstruction are well aware, there is an urgent need to develop a systematic and refined land use monitoring system. Already, promising initiatives have been taken in that direction involving several donors and research

FIGURE 8.8
Aceh. Entire Google Earth Mosaic

Source: Google Earth, 2006.

centres such as the Multi-Donor Fund for Aceh and Nias, the Aceh Forest and Environment Project (AFEP), the Leuser Foundation and the World Agroforestry Centre.

It can only be hoped that this monitoring, which the Acehnese people should be trained to handle, will provide the Acehnese, before it is too late, with key tools for protecting, rehabilitating and using productively and sustainably their exceptional forest heritage. Time is obviously of the essence.[38]

FIGURE 8.9
Southern Aceh. Google Earth Focus

Source: Google Earth, 2006.

Notes

1. This paper was prepared in the context of the Challenges of the Agrarian Transition in Southeast Asia (ChATSEA) research project, supported financially by the Social Sciences and Humanities Research Council of Canada (SSHRC).
2. "Comme beaucoup de toponymes en Asie du Sud-Est ont une étymologie d'origine végétale, il n'y a pas absurdité à penser avec W. Marsden, qu'il s'agit là d'un nom de plante, mais nous n'en avons pas la certitude"; Denys Lombard, *Le Sultanat d'Atjéh au temps d'Iskandar Muda 1607–1636* (Paris: École Française d'Extrême-Orient, 1967), p. 11. In a footnote on the same page, Lombard quotes a French-language translation of Marsden's *History of Sumatra* in the following manner: "Les Malais prétendent qu'il a ainsi été nommé d'une espèce d'arbre appelé Achi, qui lui est particulier" (*MarsHistSum* II, p. 218, note).

3. "... et tout ce que produisent les animaux de la jungle: le bézoard et le musc, le miel et la cire; et tout ce que produisent les arbres de la forêt: le camphre et l'encens, le blanc et le noir, le bois d'aigle et l'aloès et le sandal, la poix et le piment et le poivre long et tous les autres produits qui viennent des mines de la terre et des arbres de la forêt" (in Denys Lombard, 'Le sultanat d'Atjéh', p. 62).

4. Lombard, 'Le Sultanat d'Atjéh', p. 110.

5. Lombard explains how elephants were used as ramparts against the landing of enemy forces and were even embarked on ships to assist in land assaults carried out by Acehnese forces. Elephants were also employed as "work horses" in Iskandar Muda's very efficient shipyards (Lombard, 'Le Sultanat d'Atjéh', pp. 45, 81, 86, 88, 96). On the importance of elephants in seventeenth-century Aceh, see also Anthony Reid, *An Indonesian Frontier: Acehnese and Other Histories of Sumatra* (Singapore: National University of Singapore Press, 2005), particularly chapter 6, entitled "Elephants and Water in the Feasting of Seventeenth-Century Aceh".

6. With reference to the size of Aceh's territory, two types of figures are found in official sources. One fluctuates around 55,400 km² — the most frequent figure being 55,392 km² — and the other around 57,400 km². In this case the most frequent figure is 57,365 km². These are repeated throughout the literature, without anyone bothering to explain the rather substantial difference between them. The former presumably refers to mainland Aceh, while the second probably includes some if not all the surrounding islands, such as Breueh and We (Sabang), located off the north coast and (more importantly) off the southwest coast, Simeulue (extending over more than 1,800 km²) and the Pulau Banyak archipelago.

7. Charles A. Fisher, *South-east Asia: A Social, Economic and Political Geography,* 2nd ed. (London: Methuen, 1966), p. 217.

8. We have been unable to find exact figures concerning the size of the Leuser Ecosystem. Most authors refer to "2,600,000 hectares" or "26,000 km²", others even "27,000 km²". Instead we relied on our own GIS calculations carried out on the maps that we used. The results were 25,890 km² for the entire ecosystem and 3,333 km² for the section extending into North Sumatra province. Hence this section corresponds to a proportion of 12.87 per cent.

9. Leuser International Foundation (LIF), "Conserving the Leuser Ecosystem", Leuser International Foundation <http://www.leuserfoundation.org> (accessed 19 May 2009).

10. With reference to the size of the park, figures given in different sources vary significantly around 900,000 hectares (9,000 km²). The exact figure provided in a recent and carefully put together report is 862,975 hectares; see IUCN, WCPA and UNESCO, *Report on the IUCN-UNESCO World Heritage Monitoring Mission to the Tropical Rainforest Heritage of Sumatra* (Indonesia: International Union for the Conservation of Nature, United Nations Education and Social Commission, 2006), p. 34. According to our calculations — obtained through

the methods mentioned in the preceding footnote — the figure is slightly over 9,000 km².

11. IUCN/WCPA/UNESCO, *Report on the IUCN-UNESCO*, p. 30.

12. IUCN/WCPA/UNESCO, *Report on the IUCN-UNESCO*, pp. 31–32.

13. On the richness of Sumatra's and particularly Aceh's forest heritage, see Anthony J. Whitten, Sengli J. Damanik, Jazanul Anwar and Nazaruddin Hisyan, *The Ecology of Sumatra* (Yogyakarta: Gadjah Mada University Press, 1987), p. 31.

14. Frédéric Durand, *Les Forêts en Asie du Sud-Est. Recul et Exploitation. Le cas de l'Indonésie* (Paris: l'Harmattan, 1994).

15. That image has appeared on various other web sites. See <http://www.orangutanssos.org/images/forest_destruction> (accessed 19 May 2009).

16. Durand, *Les Forêts en Asie*; Forest Watch Indonesia and Global Forest Watch (FWI/GFW), *The State of the Forest: Indonesia* (Bogor, Indonesia, and Washington, DC: Forest Watch Indonesia and Global Forest Watch, 2002).

17. FWI/GFW, "State of the Forest".

18. Ibid., p. 81 (Annex table 1).

19. T.C. Whitmore, *An Introduction to Tropical Rain Forests* (Oxford: Oxford University Press, 1991), p. 168.

20. These maps are drawn from T. C. Whitmore, *Tropical Rain Forests of the Far East*, 2nd ed. (Toronto: Oxford University Press, 1984); Yves Laumonier et al., *Sumatra Sud, Carte du Tapis Végétal et des Conditions écologiques* (Toulouse: Institut de la carte Internationale du Tapis Végétal/SEAMEO-BIOTROP, 1986) 1: 1 000 000; Yves Laumonier et al., *Sumatra Centre, Carte du Tapis Végétal et des Conditions écologiques* (Toulouse: Institut de la carte Internationale du tapis végétal/SEAMEO-BIOTROP, 1983) 1: 1 000 000; Yves Laumonier et al., *Sumatra Nord, Carte du Tapis Végétal et des Conditions écologiques* (Toulouse: Institut de la Carte Internationale du Tapis Végétal/SEAMEO-BIOTROP, 1986), 1: 1 000 000.

21. These included the *World Atlas of Agriculture* (Novara: Instituto Geographico de Agostini, 1969); *Weltforstat Atlas* (Hamburg: Éditions Paul Parey, 1971); T.C. Whitmore, "Rain Forests of the Far East", in *The Last Rain Forests: A World Conservation Atlas*, edited by Mark Collins (New York: Oxford University Press, 1991); M. Collins, J. A. Sayer and T.C. Whitmore (eds.), *The Conservation Atlas of Tropical Forests: Asia and the Pacific* (New York: IUCN, 1991).

22. Rodolphe De Koninck, David Gibbons and Ibrahim Hasan, *The Green Revolution: Methods and Techniques of Assessment. A Handbook of a Study in Regions of Malaysia and Indonesia* (Québec: Département de Géographie, Université Laval, n° 7, 1977); David Gibbons, Rodolphe De Koninck and Ibrahim Hasan, *Agricultural Modernization, Poverty and Inequality: The Distributional Impact of the Green Revolution in Regions of Malaysia and Indonesia* (London: Saxon House, 1980).

23. The exact figures are 43,960 km² over 54,310 km², hence 80.94 per cent. The forest domain represented on this map includes mangroves.

24. In 1971, a giant natural gas field was discovered near Arun, in the district of Aceh Utara. Huge investments followed for extraction of the gas and the construction of a liquefied natural gas (LNG) plant in nearby Lhoksukon, under a production-sharing contract between Pertamina, the national oil company, and Mobil Oil and a Japanese LNG company. As other fields were discovered, including one offshore, massive export of oil and especially LNG began in 1978.

25. In December 1976, the Gerakan Aceh Merdeka (GAM), or Free Aceh Movement, issued a Declaration of Independence. Its leader, Hasan di Tiro, and major followers went into underground activity. Thus began a low-key insurgency, which was to frequently flare into open violence, particularly from the late 1980s onwards, between the insurgents and the Indonesian military. It lasted for nearly twenty-nine years until a ceasefire was signed in July 2005.

26. Damien Kingsbury and Lesley McCulloh, "Military Business in Aceh," in *Verandah of Violence*, edited by Anthony Reid (Singapore: Singapore University Press, 2006), p. 209.

27. Rodolphe De Koninck and Steve Déry, "Agricultural Expansion as a Tool of Population Redistribution in Southeast Asia," *Journal of Southeast Asian Studies* 28, no. 1 (1997): 1–26.

28. Down to Earth, "Aceh: Ecological War Zone", *Down to Earth* 47 (Nov. 2000).

29. FWI/GFW, "State of the Forest", p. 21.

30. Ladia Galaska is a rather complex acronym for Lautan Hindia-Gayo-Alas-Selat Melaka, meaning Indian Ocean-Gayo-Alas-Malacca Strait. The Gayo and the Alas are two indigenous groups, Alas also being the name of a major river which flows from the slopes of Mount Leuser, through the heart of the highlands and into the Indian Ocean.

31. See Down to Earth, "Ladia Galaska Road Network: Construction Continues, Controversy Rages," *Down to Earth* 62 (Aug. 2004); and several articles in *Tempo*, for example in March 2004.

32. We have actually been unable to find a single map representing the route of the highway with some degree of accuracy. In order to represent it on a map, as we did in Figure 8.2, we pieced together the highway's route, thanks to various references mentioning the towns it was to link. These are Meulaboh, on the Indian Ocean, and then Simpang Puet, Jeuram, Lhok Seumot, Ceulala, Takengon, Blangkejeren, Pinding, Lokop and, finally, Peureulak on the Straits of Malacca, some thirty-five kilometres north of Langsa.

33. Eye on Aceh, "Aceh: Logging a Conflict Zone," *Eye on Aceh* (Oct. 2004), p. 3 <http:// www.acheh-eye.org>.

34. The statistics used by Eye on Aceh are drawn from FWI/GFW, "State of the Forest"; the Ministry of Forestry (2000) and SKEPHI (2004), an Indonesian forestry NGO ("Eye on Aceh", p. 3).

35. World Wildlife Fund (WWF), *Timber for Aceh*, WWF (Mar. 2005) <http://wwf. org.au/publications/WWFTimberForAceh/> (accessed 19 May 2009).

36. *Antara News* and News.com.au, 23 December 2006; UNICEF, 27 December 2006.

37. This mosaic was carefully assembled from images available on the Google Earth website and the consultation of a number of other sites and maps. No dates are provided concerning such images, but Google Earth guarantees that they cannot be more than three years old. Hence, this figure represents the situation at any date since early 2004, most likely at some point in 2005 or 2006.

38. It is heartening to see that the Acehnese government headed by Governor Irwandi has placed a special emphasis on reducing illegal logging and promoting numerous "green" campaigns. It is hoped that this work will continue to be expanded.

PART II
Conflict Resolution

9

MANAGING RISK
Aceh, the Helsinki Accords and Indonesia's Democratic Development

Michael Morfit

INTRODUCTION: A SEASON OF SURPRISES

The final months of 2006 and early months of 2007 were marked by a series of surprises in Aceh, as the province continued a political evolution that began with the signing of the historic Helsinki Memorandum of Understanding (MoU) on 15 August 2005.[1] The first surprise was that the political campaign and elections were relatively peaceful. This was welcomed by all, but not necessarily expected by most. The second surprise was the election results. Contrary to pre-election polls, Irwandi Jusuf and his running mate, Muhammad Nazar, won nearly thirty-nine per cent of the votes, exceeding the vote threshold in order to avoid a run-off. The third surprise was the response of key stakeholders to these results. There were no cries of alarm or very negative comments from the Indonesian national government, parliament or press. Nationalistic political and military leaders who were extremely agitated over the Helsinki accords during the June–August 2005 period were also largely silent or pragmatic.

Approximately eighteen months after the signing of the Helsinki MoU, Aceh successfully negotiated key milestones and entered a new phase. Reaching almost any agreement in Helsinki was remarkable enough, but Helsinki seems to have achieved something even more unusual: a negotiated peace settlement that has taken hold, launching a new era in Aceh's political life.

This chapter examines what made the success in Helsinki possible, and how this established the foundation for the subsequent achievements in Aceh itself. It analyses three distinct but related questions concerning the Helsinki agreement:

1. The key factors that account for the success of the Helsinki negotiations
2. The path to the Helsinki MoU, which laid the foundation for subsequent, successful local government elections
3. The implications for Indonesia's democratic development and some of the lessons for the future.

There have already been some excellent descriptions of the events leading up to the Helsinki negotiations and step-by-step narratives of the negotiations themselves as they unfolded over the January–August period.[2] More recent publications have given in-depth and personal perspectives, drawing on the accounts of those participating in the negotiations.[3] This analysis builds on these earlier efforts. But it is less a chronology of *what* happened than an effort to get inside the minds of those who laboured to reach an agreement in Helsinki and worked to keep the peace process on track afterwards. Drawing largely from direct personal interviews with key participants from all sides, including GAM, government, mediators and advisers, this chapter tries to illuminate the objectives, strategies, risks and benefits, as these were understood by each side at the outset; how these evolved in response to the Helsinki dialogue; and how they have been shaped by external forces and events since Helsinki.

The chapter begins with a summary of the often-cited explanations for the initiation and ultimate success of the negotiations, arguing that they are plausible but incomplete and ultimately unsatisfactory. It then examines the eighteen-month period prior to the first round in January 2005, in an effort to illuminate the perspectives of each side going into the negotiations; it looks at how each side attempted to manage their perceived risks, as reflected in their strategies and approaches. And, the roles of individual foreigners and international agencies are examined as part of the strategies and approaches used by each side, rather than as external and independent variables.

The post-Helsinki political developments in Aceh will also be examined, particularly the responses of the national political elite to the surprising victory of the former GAM leader, Irwandi Jusuf. Finally, this chapter argues that some of the key factors that helped secure a successful outcome of the negotiations in 2005 may become liabilities in the continuing political evolution of Aceh.

In all these sections, particular attention is given to the internal dynamics and perspective of the national government in Jakarta. Less emphasis is given to GAM for several reasons. First, the issues, options and actions of the Jakarta government were far more transparent and readily accessible. The key actors were part of an increasingly vigorous democratic system; their policies and decisions could not escape public debate. In contrast, GAM was not an established national government but a political movement that included an armed insurgency. Not surprisingly, its internal processes and deliberations were far from transparent, accessible or participatory. The physical survival of the GAM leadership and their movement depended on GAM's internal discipline and unity; their experience demonstrated the necessity of moving with extreme caution with regard to any external actors. Even today, it is still not easy to penetrate the world of the GAM leadership, understand their issues and trace their internal debates. The discipline that helped hold GAM together through nearly three decades of armed struggle persisted through the Helsinki negotiations, as the GAM leadership closed ranks around agreed positions, rather than talk about differences and choices. Ironically, the provincial elections in December 2006 significantly eroded this discipline. At a key meeting in May 2006, GAM was unable to agree on a single slate of candidates. Over the ensuing months, key figures in its leadership split into different political factions to contest the December elections.[4]

More importantly, however, this chapter gives more attention to the national government, because of the great challenges they faced. In many ways, the national government in Jakarta faced a more complex environment than GAM and more wide-ranging risks. It is true that GAM relinquished their aspiration for full independence: this was a difficult and risky decision. For Jakarta, however, the path to Helsinki may have been significantly harder, and the consequences of missteps or failures greater. This is the aspect of the Helsinki process that has not yet been fully explored, yet it is essential if we are trying to understand not only how it was possible to reach agreement but also why it seems to have endured to shape the new political era in Aceh.

Finally, the perspective of the national government is given greater attention because the way in which it responded to these challenges has very significant implications for Indonesia's future political system. This is the core conclusion of the entire chapter. The success of the Helsinki negotiations was more than the resolution of a long-standing and bloody conflict. It also marked an important achievement in Indonesia's democratic governance in general, and in the assertion of civilian control over the military in particular. A narrowly focused study that examines only Aceh and not the wider political context tends to miss or obscure these wider implications and to overlook the extent to which the Helsinki MoU is a key milestone in the nation's democratic development. This may be the most far-reaching dimension of the Helsinki agreements — one that has been largely overlooked or underappreciated.

CONVENTIONAL WISDOM

This chapter departs from several explanations that constitute the prevailing conventional wisdom about the reasons for the success of the Helsinki negotiations. These core assessments have shaped popular discourse, from press coverage to political speeches and diplomatic perceptions. They are not mutually exclusive explanations, but they are distinct because each highlights a different aspect of the negotiation process and emphasizes different forces at play.

Perhaps the most commonly held assumption about the peace process is the so-called tsunami factor. It is widely claimed that everything in Aceh changed fundamentally as a result of the tsunami on 26 December 2004, forcing both sides to reconsider their positions and take advantage of significant international assistance to respond to a devastating crisis. This is why negotiations began in January 2005, after the tsunami hit Aceh and profoundly altered not only the physical and human environment, but also the political environment.[5]

A second explanation is the so-called TNI factor. There is a broad agreement that the military operations launched by the TNI, when the Cessation of Hostilities Agreement (CoHA) collapsed in May 2003, had a significant impact on the political presence, systems of financing and military capacities of GAM. Some argue that GAM became essentially a broken force militarily. Its ability to sustain the armed conflict was severely, even mortally weakened; as a result, it had no choice but to seek a negotiated settlement.[6]

A third conventional explanation is the "Kalla factor". Vice-president Jusuf Kalla was actively involved in Aceh issues long before the formal Helsinki process was launched and was prominent throughout the negotiations. His investment of time, energy and political capital was both very unusual and highly visible, and is often cited as a key factor in reaching agreement.[7]

These conventional explanations are not completely wrong, but nor are they entirely satisfactory. A closer examination of the evidence suggests that they are pieces of the puzzle, but incomplete and even misleading because they overlook the complexity of events preceding the first round of negotiations in January 2005.

The first clue that they are only part of the story emerges from a careful review of the chronology of events leading up to the first round of negotiations in January 2005. Key sources interviewed for this study agree that, by mid-December 2004, concrete plans were already well underway to convene the first round of negotiations in Helsinki. Prior to the tsunami, the head of the Crisis Management Initiative (CMI), former Finnish president Maarti Ahtisaari, was seeking confirmation from both sides on basic understandings prior to taking on the role of facilitating negotiations in Helsinki. On 23 December 2004, three days *before* the tsunami unexpectedly hit Aceh, the Government of Indonesia confirmed that its delegation would be led by Minister of Justice Hamid Awaluddin.[8] Plans for Ahtisaari to meet face to face with the GAM leadership were also already far advanced and postponed only because of the tsunami. The advanced stage of preparations for negotiations in Helsinki are alone enough to call into question the "tsunami factor" as the key reason for success in the negotiations.

The formal invitations were themselves the culmination of continuous efforts to establish a foundation for direct negotiations that stretched back for at least eighteen months to June 2003. Interviews with key participants illuminate the extent to which this earlier period was critical to what was later achieved in Helsinki. These earlier efforts laid the foundation for the Helsinki talks, but they were largely managed separately from the TNI military operations. In addition, while Kalla was a key figure in these preparations, he was not pursuing an independent foreign policy but enjoyed critical support from President Susilo Bambang Yudhoyono (SBY).

In short, the road to Helsinki started long before December 2004. A more detailed examination of the dynamics of the negotiations during this period reveal how each side managed its own complicated web of constituencies and forces. In turn, this helps explain why the MoU has taken hold, how it helped shape the environment for the recent successful elections and why it is a significant milestone in Indonesia's democratic development.

MANAGING RISKS

No Simple Asymmetry

In assessing the positions and capacities of the two negotiating parties at the end of 2004, the initial impression is one of a clear asymmetry between the national government and GAM as well as of a highly unequal contest between a large and powerful state enjoying broad international support and a tough but severely weakened and struggling armed insurgency under increasing military pressure. GAM was also unable to mobilize significant international opposition to the government's military operations following the collapse of the Cessation of Hostilities Agreement (CoHA) and the imposition of martial law in May 2003. GAM may have been too confident and overplayed its hand. In the process, it appears to have irritated or disappointed members of the international community who had persuaded the Government of Indonesia to seek a peaceful settlement. As a result, the international community offered little objection to the government's decision to impose martial law and launch a military initiative.[9] Certainly, during 2003–04, GAM appeared increasingly isolated and vulnerable. Several external observers and direct participants have characterized the GAM leadership as unsophisticated and even inept in the lead-up to the Helsinki negotiations.[10]

This initial impression, of a strong national government facing a weakened insurgency, appears to support the conventional analysis that Indonesian national military (TNI) operations were very successful. It can be argued that the tsunami then gave GAM a face-saving reason to accept the realities of military defeat, relinquishing its claims for independence and accepting special autonomy. The energy, flexibility and skill of Jusuf Kalla and his hand-picked team then allowed the government to exploit their clear advantage and push negotiations to a successful conclusion.

However, looking at the earlier period from mid-2003 to January 2005, a far different, more complicated and more intriguing picture emerges. As discussed below, the evidence suggests that, in many respects, GAM was in a more straightforward and easily managed position than the government, with a more limited range of vulnerabilities and fewer risks than the national government in Jakarta.

The Pursuing and the Pursued

One indication of the complexity of the bargaining positions of GAM and the government is the extent to which both President Susilo Bambang Yudhoyono and Vice-President Jusuf Kalla invested their personal time, energy and political capital in developing, nurturing and pursuing the idea of a negotiated settlement, far in advance of the start of formal talks in January 2005. In parallel with the military operations being pursued under martial law and the civil emergency (2003–04), SBY and Kalla were at the heart of a distinct — at times almost independent — stream of manoeuvring, trial-

and-error, informal discussions and preparation, virtually all of which were initiated by the Jakarta government.[11]

Kalla was at the forefront of most of these efforts. Very shortly after the collapse of CoHA and the imposition of martial law in May 2003, Kalla took the initiative to propose to then President Megawati that he continue to seek ways of re-engaging GAM in a dialogue. While Megawati did not oppose this initiative, she did not enthusiastically embrace it or make it part of a coherent overall government strategy. Instead, Megawati appears to have been content to let Kalla proceed at his own pace, while at the same time giving more or less free rein to the TNI to pursue military operations. Kalla was given no written instructions or authorization, and the scope of his authority appears to have been vague. To protect his efforts, he made sure that both SBY (then coordinating minister for security and politics) and General Endriartono Sutarto (then TNI commander-in-chief) were present when his proposal was discussed with Megawati. Only when he knew he had their support for his efforts — effectively covering his flank with the TNI — did he proceed.[12]

With this vaguely defined mandate, Kalla turned to his trusted assistant, Farid Husain, charging him to find a way to talk to the GAM leadership. Over the next eighteen months, Husain (and sometimes Kalla personally) pursued opportunities for re-establishing a dialogue with GAM, which was clearly not an easy sell. There was every reason for GAM to believe that the real strategy of the government was pursuing a complete military victory, not resurrecting peace negotiations. In the eyes of the GAM leadership, the government had to meet a significant burden of proof before any negotiations could be taken seriously. Although Malik Mahmud has acknowledged that "the existing strategies applied by both parties had caused a costly stalemate", GAM was not actively seeking negotiations.[13] Despite the damage inflicted by the TNI, GAM appeared cautious rather than desperate, sceptical about the value of talks and apparently prepared to hunker down and weather continued military pressure from the TNI. Even after the Helsinki talks got underway, GAM leaders were highly doubtful about the government's commitment and intentions. According to Nur Djuli, one of the GAM negotiating team, "at this point, we were not that serious. We were just being polite, but we did not have any expectations that new talks would have any success and we were not really committed to the process".[14]

This is not to suggest that GAM was completely indifferent to the idea of re-establishing a dialogue with Jakarta. Nonetheless, the overall picture that emerges is somewhat surprising. In May 2003, the government had walked away from the CoHA and launched renewed military action. But almost immediately, both in and out of government, Kalla and SBY were in one way or another directly pursuing GAM, trying to establish contact, open communications, credibility and a basis for negotiations. In contrast to the intense engagement of the two most senior Indonesian leaders, GAM leaders were distant, cautious and sceptical. They had to be pursued and persuaded to go down the road to Helsinki.

More Complex Challenges for the Government than GAM

The close attention that SBY and Kalla gave to pursuing GAM is perhaps less surprising when we understand that it was the government in Jakarta — not GAM — that faced the most complex challenges, incurred more diverse risks and, in many respects, had to go the furthest distance in order to bring the negotiations to a successful conclusion. This is why negotiating with GAM was not a simple or routine task that could be easily delegated. It was a highly sensitive responsibility, and success was far from assured. It is understandable, therefore, that it commanded such high-level attention.

The starting point is the recognition that any successful negotiation requires credible partners. Each party has to be convinced that, if agreement is reached, the other side will be able to deliver what it has promised. Regardless of the specific issues and details, no negotiation can succeed or endure if one or both of the parties is unwilling, uncertain or incapable of meeting its agreed obligations.

This basic requirement is precisely where GAM was in a more advantageous position than the government in Jakarta. It is true that GAM faced a host of adverse factors, including the pressures of the TNI; disruption of GAM political and military structures in Aceh; the resulting shrinking of its revenues; the extraordinary logistical challenges in holding a geographically dispersed movement together; increasing international isolation; and a relatively insular and inexperienced negotiating team. Yet, despite these pressures, GAM consistently demonstrated "remarkable discipline and remarkable strength in their chain of command, as well as a willingness to change tactics".[15] Farid Husain noted the "strong collective leadership of GAM" and commented that "we could not isolate one from another, and build a separate relationship with only one of them".[16] The government made several attempts to marginalize the GAM leadership in Sweden and to deal directly with field commanders. All were uniformly unsuccessful.

GAM's extraordinary discipline can in part be explained by its single-minded goal (independence for Aceh), with no specific social programme or agenda and none of the institutional checks and balances of an emerging democratic system. This made it relatively easier to ensure the coherence and discipline of the movement. In contrast to GAM, the national government in Jakarta faced much more complex and wide-ranging issues, competing objectives, diverse constituencies and strongly vested interests. These included issues similar to Aceh, such as separatist movements in Papua and continuing communal conflict in Poso and Ambon, as well the ongoing, high-profile challenges of national government, ranging from the continued threats of terrorism and avian flu to economic reform and promoting growth, judicial reform and anti-corruption measures.

Neither SBY nor Kalla were in any doubt about the challenges they faced from within their own system of government. SBY believed that the most important opposition to his whole approach came from some very senior elements of the TNI and from vocal nationalist politicians in the DPR (national parliament). In his view, both groups were "very rigid" — unwilling to compromise and highly suspicious of or even opposed to any negotiations with GAM.[17] They knew that these groups both had wide-reaching and powerful networks in the military, civil service, political parties and private sector, and that they had demonstrated the capacity to undermine, erode or sabotage any agreements that they did not support. Previous efforts to secure a sustainable peace arrangement during the Humanitarian Pause and CoHA were consistently criticized by nationalist politicians in the parliament and actively undermined by military commanders in the field.[18]

Ironically, Indonesia's democratic transition has made managing opposition from these two groups more complicated. Rather than being a key reason why the Helsinki negotiations succeeded, a vibrant democratic system has created many more challenges in pursuing a policy that has strong opposition from key stakeholders.[19] With a free and active press, expanding and lively civil society, vibrant (if undisciplined) political parties and a vocal (if not always mature) national parliament, any and all policies are now subject to increased demands for transparency, public debate and accountability.[20]

The failure of earlier peace efforts had something to do with GAM, but it also had a great deal to do with the failure of previous administrations in Jakarta to master this complex, evolving and unruly political environment. Previous administrations were unable to articulate a coherent approach to Aceh, forge agreement among key stakeholders and enforce discipline within their own ranks. To succeed where previous administrations had failed, SBY and Kalla had to manage these powerful constituencies in Jakarta before they could engage GAM as a credible, constructive and coherent negotiating partner.

Higher Stakes

In addition to a more complex and challenging political and institutional environment, the national government also risked greater consequences. From the perspective of the GAM leadership, failure in Helsinki would almost certainly have resulted in intensified military pressure from the TNI. It would have been very difficult, but GAM had already demonstrated both its discipline and resilience. The GAM leaders acknowledged that they could be forced to retreat, but they were confident that they could not be forced to surrender.

For the new SBY/Kalla administration, however, the damage from failure would have been much more extensive. SBY's support for the peace process during the CoHA period had already generated strong criticism and increasing pressure from both senior TNI officers and nationalistic politicians.[21] In reviving the idea of negotiations with GAM, his administration was again taking on the same powerful vested interests that had thwarted earlier administrations. Failure would have reinforced a pattern of undisciplined and unfocused policy processes, with wide latitude for covert influence and subterfuge of declared government policies by the military and ultra-nationalists.

In a sense, this might be described as merely the risk of a return to "business as usual", continuing the fitful and wavering policy process of the Gus Dur and Megawati administrations. But SBY and Kalla had come to office with a pledge of firm leadership, new directions and progress. "Business as usual" would have been a significant step back from their stated objectives. In addition, SBY's ability to pursue other priorities — from separatist movements in Papua to governance reform and anti-corruption — had to confront many of these same vested interests at different barricades; his ability to make progress on these other fronts would have been significantly weakened from the outset of his administration.

There were also strong fiscal incentives to seek a negotiated settlement and end the military conflict in Aceh. SBY had long believed that the "conflict had gone on too long; there were too many victims on both sides. And it was expensive, costing us about $130 million per year in security operations".[22] Interviews with TNI staff and families by Kirsten Shulze revealed the TNI support systems were severely strained. Lack of sophisticated navigation equipment prevented the rapid evacuation of wounded troops, with the result that simple wounds often resulted in fatalities. Resolving the conflict would ease the government's fiscal burdens and the human costs associated with a military system under significant stress.

Failure to negotiate a settlement in Aceh would undermine efforts to project an image of security and stability within Indonesia, a critical requirement in attracting private investment, strengthening economic growth and meeting one of SBY's key election promises of improving the economic welfare of the Indonesian people in all parts of the country. Prolonged conflict would also undermine the clear aspiration of the SBY administration to raise Indonesia's international profile, revive Indonesian leadership within the region, and advance the administration's goals for Indonesia to be accepted as a neutral, reasonable, steady and reliable international partner.

Personal Convictions and Institutional Benefits

In the critical period of preparation in advance of the Helsinki talks, the conflict in Aceh was not a central political issue in Indonesia. There were, after all, a lot of other issues that dominated national political discourse during 2003–05. The political ambitions of both SBY and Kalla were not going to be significantly advanced by a negotiated settlement in Aceh. Public opinion polls showed strong support for Megawati when she imposed martial law in Aceh and launched the largest military operation in Indonesian history in May 2003.[23] Parliamentary opposition to military action in Aceh was nonexistent, and the press was neither aggressive nor inquiring, being content largely to report the campaign from the TNI perspective.

Given this context, personal pride may be part of the explanation for the willingness of both men to invest such significant time and energy on this issue. There was also an apparently genuine and firm conviction that any attempt at a purely military solution in Aceh was doomed to failure, that the conflict could be resolved only on a sustained basis through negotiations. SBY's direct involvement in Aceh issues goes back as far as Gus Dur's administration and included the period when he served as coordinating minister for Political, Security and Defense Affairs. Although not the leader of the government efforts, he was closely involved with the negotiations that led to the CoHA and remained a defender of that agreement, even in the face of public complaints from senior TNI commanders that GAM was using the agreement only to expand its military and political presence.[24] His strong view that no purely military solution was possible was directly opposed to confident predictions from some TNI leaders that the military operations launched in May 2003 would permanently crush GAM and resolve the conflict in Aceh. He had always been in the minority.[25] SBY thus had a personal stake in disproving that basic position.

Jusuf Kalla shared SBY's strong view that only a negotiated settlement could yield sustained peace. He took understandable pride in his leading role in the settlement of religious conflicts in Central Sulawesi and Ambon when he served as coordinating minister for People's Welfare under Megawati. His success in resolving these difficult and bloody conflicts could be followed by a third achievement in Aceh. Kalla already had secured an informal mandate from Megawati to seek ways of re-establishing dialogue with GAM. He is not someone who is easily discouraged: he knew that if he persevered and found the path to success in Helsinki, it would enhance his prestige and standing even further.

These personal considerations are almost certainly important. However, resolving the conflict in Aceh was also important for reasons that went beyond the immediate confrontation with GAM. Aceh was the venue for a clash that had broader national significance. How the government articulated, developed and implemented its policies with regard to Aceh would have great impact on the prestige, power and prospects of the new SBY/Kalla administration. It also had direct implications for Indonesia's transition towards more open, accountable and effective government. Success in reaching agreement in Helsinki would demonstrate the resilience and flexibility of the national system in accommodating regional differences yet maintaining national unity. Bringing potential spoilers (especially in the TNI) on board and limiting opportunities for undermining the negotiations and subsequent implementation of the MoU was critically important for continued progress in establishing civilian control over the military. Enforcing some coherence and focus to the government's approach was important in strengthening the ability of the executive to formulate and implement policies without informal subversion by dissident factions.

MANAGING RISK

The preceding discussion suggests that the endurance of the Helsinki MoU can be explained by the strengths, weaknesses, opportunities and risks as these were perceived by the national government in Jakarta. This section examines the way in which SBY and Kalla assessed and managed this environment. It focuses on how the administration tried to confront, convince, co-opt, manage or marginalize the key opponents from within their own side. This not only made reaching an agreement possible in August 2005, it also laid the foundation for the resilience of the MoU and the successful transition of the local government elections.

"Political Umbrella": Spending Political Capital for a Common Vision

Even before their inauguration, SBY and Kalla shared a common vision in approaching the Aceh problem. According to SBY, "I believed very strongly that a military solution could not solve the problem permanently and conclusively. We have 50 years of experience to prove this, not only in Aceh".[26] Similarly, Kalla had a strong philosophical commitment to dialogue as a means of resolving disputes. The clear and strong commitment of SBY and Kalla provided the often elusive "political will" that is a critical ingredient of successful political change. SBY describes his commitment as providing the crucial "political umbrella" that protected the work of the negotiators. This policy coherence at the topmost level of the government was a huge advance over previous administrations. Indeed, it could be argued that lack of unity on the Government of Indonesia side, characterized by continuing opposition to negotiations and covert subversion of efforts to find a peaceful resolution, was the fundamental cause for the failure of earlier negotiations.[27] SBY did not have to worry about a conflicting policy coming from within his own staff or a competing political centre in the Office of the Vice President. SBY and Kalla understood that the challenge was not simply to restart talks, but to be able to defend and implement them in the face of what they knew would be strong internal opposition.

The SBY-Kalla Team

Given their common commitment to finding a negotiated settlement, SBY described his role as setting the objectives and developing the broad strategy. This included careful calibration of competing demands and interests. According to SBY, these included three distinct groups of stakeholders. The first was the local population directly affected by the situation in Aceh: "I had to demonstrate that we were making progress in reducing the armed conflict and turning the province into some kind of normal state". Second, he had to assess and manage powerful national constituencies that would be strongly opposed to any settlement they regarded as a betrayal of fundamental nationalist principles. These included political parties, the current TNI leadership and retired TNI commanders who retained informal links and commanded loyalties within the TNI. Third, SBY had an eye on the international community. With the very large support of the international community for post-tsunami reconstruction, Aceh was very much in the news. However, he also felt that the international community was quietly watching how his administration would resolve the conflict with GAM and the longer-term implications for political developments in Indonesia.[28]

Effectively managing these three different groups of stakeholders — local, national and international — was key to reaching agreement in Helsinki. "I had to be engaged on all three fronts", SBY has commented. This was a lesson he believed he had learned from the past, when Megawati's failure to keep fully engaged in managing key stakeholders meant that there was not the political support and coverage required.[29]

While SBY had his eye on this larger strategic picture, he delegated to Kalla the responsibility for overseeing and managing the Helsinki negotiations on a day-to-day basis. If SBY provided the political umbrella, then Kalla is widely credited with driving the negotiations forward. This included closely monitoring ongoing talks, identifying key issues, examining technical questions, identifying options, reviewing and amending texts, and recommending final approvals by the president. Kalla was deeply immersed in the details of the negotiations, often personally drafting analyses of different options or background papers on key issues.[30] As the negotiations reached their final stages, Kalla closely reviewed, commented on and amended each of the progressive drafts of the MoU between the two sides.

Kalla's direct day-to-day engagement was highly visible, frequently putting him (or those on the Indonesian delegation who were seen as "Kalla's men") in the spotlight, while the engagement of SBY was less evident. SBY argues that this was deliberate: "I could not always be the forefront. Sometimes I had to be in the background". He let others take the lead, preserving political capital until it was really needed.[31] This gave the administration greater flexibility and more depth in dealing with potential opponents.

The contrasting political bases and personal networks of the two men, sometimes cited as a source of tension or conflict within the administration, were key to their effectiveness as a team in resolving the Aceh conflict. Their individual differences greatly increased their combined reach across the spectrum of stakeholders and their ability to contain and manage potential spoilers. Each had their strengths in distinct but equally necessary constituencies. In the words of Jusuf Kalla, "there was a division of labour. SBY took care of the TNI and I took care of the political parties, as well as the technical issues".[32] Neither man would have been able to accomplish the task of co-opting and managing both these potential spoilers on their own. Together they were able to manage all key constituencies and to bring an unusual degree of coherence and discipline to the government's approach.

There is another sense in which the contrasting personal styles of the two men were complementary and effective. SBY is well known for his extremely measured, cautious and deliberative style. His responses to questions are typically slow, thoughtful and considered. Kalla, in contrast, is quick and clipped, often leaping from one thought to another in an effort to reach a conclusion. One can imagine that a conservative TNI commander or ultra-nationalist politician might be a little nervous about what Kalla might agree to. SBY, in contrast, presents a calm, balanced and thoughtful image, with a record of cautiously balancing competing interests and calculating how fast and how far he can go. For many who were highly suspicious of any negotiations with GAM, this had to offer a reassuring sense of restraint and balance.

Bringing the TNI on Board

All analyses of earlier efforts to resolve the Aceh conflict in the post-Suharto era agree that ensuring TNI support for a political settlement was a challenge that previous administrations had failed to meet.[33] In general, key elements of the TNI leadership seemed to be opposed to anything other than a complete military defeat of GAM. For example, "when *Operasi Terpadu* was launched [in May 2003], TNI commander-in-Chief General Sutarto ordered his troops to 'destroy GAM forces down to their roots' by 'finishing off, killing, those who still engage in armed resistance'".[34]

There are various explanations for this often strongly stated position, ranging from principled nationalism and an unwillingness to compromise the territorial integrity of the Indonesian state to craven economic self-interest fuelled by corruption.[35] Whatever the explanation, the TNI was clearly a force that had enjoyed considerable, if sometimes shifting, latitude in earlier administrations, often not hesitating to dissent from declared government policies. Somehow, the TNI leadership had to be brought on board.

SBY moved remarkably quickly in the very early weeks of his administration to address this challenge. In the final days of her administration, after she had already lost her re-election bid to SBY, President Megawati had sent a letter to the DPR, nominating General Ryamizard Ryacudu (then the army chief of staff) as the next TNI commander-in-chief. Ryacudu was widely regarded as a "natural" choice and had strong support from within the TNI. Nonetheless, SBY recalled the letter of nomination and instead proposed extending the tenure of the incumbent, General Endriartono Sutarto. This was a critically important decision. Ryacudu was "the most vocal representative of the anti-reform wing of the armed forces.... As army chief of staff, he was not only a visible symbol for the military's reluctance to further reform, but he also had the power to influence the outcome of important policy processes. In early 2003, Ryacudu had belonged to the fiercest opponents of the Aceh peace process, and many believed that he played a major role in its failure".[36] Various key government participants in the Helsinki process have stated categorically that it would have been impossible to reach an agreement with GAM had Megawati's proposed appointment of Ryacudu gone forward.[37]

In contrast, General Endriartono Sutarto, although not a radical reformer, was trusted by SBY as someone "able to look at the wider picture" and "a supporter of the peace process".[38] Those most closely involved in the negotiations felt that Sutarto had a sophisticated world view, finely tuned political instincts and was less doctrinaire, with a pragmatic willingness to adapt and adjust.[39] SBY used Sutarto to help pull the TNI into line with his policies. When Ryacudu continued to speak out against negotiations with GAM, SBY asked Sutarto to "control statements of the military and to follow government policies".[40] Sutarto has confirmed that "I told the TNI leadership that I don't want any senior officers talking out against government policies. If you want to oppose government policies, then you must leave the TNI".[41] SBY followed this up in March 2005 by delivering the same message at TNI headquarters, when he attended the ceremony for the installation of new service chiefs.[42]

SBY claims that this was a calculated risk, but he was confident of the outcome. Shortly after the fall of Suharto, when public opinion was demanding significant military reforms, SBY distributed a questionnaire to senior officers to get their views on how the military should respond. He learned that about 60 per cent supported reforms, but felt they must be gradual and well controlled. These "moderates" were the largest group, but not outspoken or high-profile, and this is where SBY placed his own views. About 25 per cent were hostile to any reforms, and about 15 per cent wanted faster reforms. The younger generation of officers (that is, colonels and below) were less political, more flexible and open-minded, suggesting that the time clearly favoured some kind of reform process.[43] SBY recalls, "I knew my audience" and "I had it mapped out," although "I had to calculate carefully because I knew that I needed their support".[44]

Establishing a Clear Policy and Approach

SBY argues that there was a lack of coordination and direction in the government's Aceh policy in the Gus Dur administration. Under Megawati, there was better coordination and clearer objectives, but still no clear strategy that had the commitment and full support of the president. In his administration, SBY felt it was critical to establish a clear framework or "comprehensive solution" that could help orient, direct and coordinate the work of the negotiating team.[45]

The negotiating team was told that there were two fundamental requirements for any agreement: first, GAM must accept the principle of the territorial integrity of the Indonesian state (Negara Kesatuan Republik

Indonesia, or NKRI); second, any agreement must be consistent with and in the framework of the Indonesian constitution. According to SBY, this was the *harga mati*, or fixed bottom line. All else was negotiable, and the negotiating team had great latitude as long as they remained within this framework.[46] Either working on his own initiative or with the explicit encouragement and approval of SBY, Jusuf Kalla developed the basic government approach in a memo formally transmitted to the president on 9 January 2006. Once SBY blessed this statement, it provided a common reference point for the team and was the basis for SBY's public announcement on 10 January that negotiations with GAM would begin at the end of the month. In the words of Kalla, "this provided the clear framework that we did not have for CoHA. This was the basic framework for the whole negotiation. We did not really stray from this".[47]

Arguably, these core principles were not original to the SBY administration. President Megawati had similarly emphasized the maintenance of national unity as her top national goal.[48] Although broadly stated, the framework was based on a careful calculation of what SBY and Kalla believed they could ultimately sell to key stakeholders in Jakarta, such as the TNI leadership and leaders of political parties represented in the DPR. These were, of course, the very same constituencies faced by former administrations. So, in one sense, not much had changed. What was new, however, was the clear expectation that the path to achieving these goals had to be one of negotiation and political compromise, and that pursuing that policy had the full and unequivocal support of both the president and vice-president.

This framework did not resolve all issues, of course. It was too general to be useful in refining the government's response to difficult but critically important questions (most notably, the possibility of creating locally based political parties). Nonetheless, the fact that these other key issues were clearly seen by the Government of Indonesia as open to discussion and negotiation helped to orient and guide the work of the negotiating team.

Working out of Channels

Having committed themselves fully to the search for a peaceful resolution of the Aceh conflict, and with a clear framework for the government delegation, SBY and Kalla then pulled together an unusual team to meet with GAM in Helsinki. They sidestepped the formal government bureaucracy and deliberately appointed an ad hoc team comprised of hand-picked individuals who enjoyed their personal confidence, could talk directly and frankly to the president and vice-president and brought specific skills and experience to the table. The three core members of the team were selected for these personal qualities rather than because of their formal rank or government departments. Hamid Awaluddin (minister of Justice and Human Rights) was the delegation leader and had worked closely with Jusuf Kalla on the Poso and Ambon conflicts. Sofyan Djalil (minister of Communications) is Acehnese by origin and was in part selected because of a desire to establish direct and personal connections with the GAM delegation. Farid Husain (a director-general in the Department of Health) had been pursuing the GAM leadership for over eighteen months and was valued because of his affable personality and skills as a facilitator.

The fact that none of the leading members of the Indonesian team were Javanese was also a deliberate consideration. It was an effort to be sensitive to GAM's aversion to "Javanese colonization" of Aceh.[49] In addition, the Indonesian delegation deliberately avoided bureaucratic formalities. Once negotiations got underway, "we were determined to deal with GAM with dignity, not the take-it-or-leave-it attitude of past negotiations"; "If we had approached this in the normal way, as if we were involved in a normal diplomatic negotiation, with all the formality and structures of the bureaucracy, we probably would have found it difficult to reach agreement. It was necessary for us to have great flexibility, although we consulted regularly with Jakarta".[50]

There were some concessions to a more conventional representation on the delegation. Admiral Adi Sudjipto Widodo, coordinating minister for Political and Security Affairs, participated only in the first two rounds of negotiations in Helsinki. He says that once he was satisfied that the basic framework established by SBY and Kalla was firmly established and not threatened by the negotiations, he felt comfortable excusing himself from subsequent sessions.[51] His deputy, Usman Basja, was then left to represent his interests. Only in the final few rounds, when issues of disarmament, demobilization and re-integration were being discussed, did TNI representatives join the delegation. Wesaka Pudja from the Ministry of Foreign Affairs was present throughout as the note-taker, providing daily summaries to Jakarta, but seems not to have played a major role in any of the decision-making.

There appears to have been remarkably little in the way of bureaucratic involvement or support for this team. Most of the discussions were by telephone between the government delegation in Helsinki and Kalla in Jakarta. Before departure for Helsinki and after returning to Jakarta, the team would routinely meet with the president to brief him and discuss plans for the next round. Beyond this, there appears to have been little in the way of a formal inter-agency process, laboriously drafted position papers or lengthy clearance process.

All this clearly indicates that SBY and Kalla deliberately reduced opportunities for bureaucratic delay, equivocation and sabotage by hand-picking the negotiating team, managing the process "out of channels" and relying on a highly informal policy process. Information flows were carefully controlled — a singular achievement in a political culture where informal networks, unauthorized leaks and rumours are commonly used to influence the political process. The resulting combination of high-level political support, maximum flexibility and relatively disciplined communications were key to the government's ability to manage its own constituencies and stakeholders, including critically important potential spoilers in the TNI.

A Fundamentally Indonesian Process

Finally, although assistance and support from outside agencies was critically important at pivotal moments, this was fundamentally an Indonesian process, driven by Indonesian actors and managed in an Indonesian manner. International institutions and foreign experts could facilitate and support, but they could not determine the outcomes. From the very earliest stages of the road to Helsinki, there was a pattern of informal and highly personal contacts; use of intermediaries and personal networks; informal and behind-the-scenes negotiations; and ad hoc, trial-and-error approaches.

INTERNATIONAL FRIENDS: INSPIRED AMATEURS AND EXPERIENCED EXPERTS

Notwithstanding the commitment and engagement of SBY and Kalla, and their success in getting GAM to the negotiating table, both the government and GAM had some help from international friends. This assistance came from an intriguing and fortuitous combination of inspired amateurs and experienced experts who were able to help narrow the gap of distrust that separated the two sides.[52]

One inspired amateur was a private Finnish citizen, Juha Christensen, who took a personal interest in trying to help the government establish some connection with the GAM leadership. Through contacts, he was able to arrange a meeting with a former high-ranking UN official and ex-president of Finland, Ahtisaari, who was also the head of CMI. After a series of phone calls, Ahtisaari agreed to meet with Christensen and Husain in Helsinki. This was the fortuitous entrance of a key figure who was later to chair the negotiations in Helsinki and help steer them to success. Both sides credit Ahtisaari's skills as an experienced diplomat, shrewd politician and forceful personality for pushing them past difficult moments and towards final agreement.[53] In the eyes of the GAM leadership, Ahtisaari brought international stature and credibility that the Henry Dunant Centre (HDC) lacked during the CoHA period. Ahtisaari's international prominence and connections gave the sceptical GAM leaders the reassurance they needed that it was at least worth listening to what the government had to say.[54]

Ahtisaari had an unusual and unique range of assets that proved critical in keeping the negotiations on track. These were personal rather than institutional and derived largely from his experience as an international civil servant and politician. First, he was a forceful and energetic chairman who was rigorous about keeping discussions focused and not hesitant about browbeating the delegations when he felt they were wandering into unfruitful topics.

Mindful of his own reputation and credibility, Ahtisaari was prepared to be tough and demanding with both sides: "I made it clear that I was doing them a favor. If both sides did not come prepared for serious negotiations, I told them I was not interested in wasting my time and energy". At various points throughout the negotiations, Ahtisaari says he "set up tests to determine if they were really committed", such as insisting that there be no leaks about the progress of negotiations and that the information made available to the press be carefully controlled.[55]

Second, he was clear from the outset about the scope and ground rules of the discussions, as well as the mandate of CMI. The negotiations would specifically not include full independence for Aceh, but they would explore options for an expansion of local government authorities beyond the established legal framework of "special autonomy". He established the basic principle that "nothing is agreed until everything is agreed". This forced both sides to look for a comprehensive settlement that included even the most difficult issues, rather than seeking quick agreement on individual issues but evading some of the core problems that separated the two sides.[56] Finally, CMI would facilitate discussions, but (unlike the HDC) it would not undertake any responsibility for monitoring implementation.[57]

Third, Ahtisaari was able to draw on an exceptionally wide personal network to bring outside resources and expertise to the negotiations. Out of deference to Ahtisaari's previous position as head of state, the Government of Finland agreed to provide the venue and logistical support. As the negotiations proceeded, Ahtisaari was able to borrow the services of a Finnish colonel to help advise on issues of disarmament, demobilization and re-integration.

He also used his connections with the EU to persuade them to send some "observers" to the final rounds of negotiations, and then to expand that involvement to EU participation in the Aceh Monitoring Mission (AMM), as discussed in Peter Feith's chapter in this volume.[58]

The engagement of the EU in monitoring the implementation of the agreement was particularly important for both sides. Notwithstanding progress on other issues, there was still strong distrust about the ability of their opposite numbers to adhere to any agreement. Both sides regarded the other as having exploited and subverted previous peace agreements, and both believed the same pattern could easily be repeated in the future. Only a robust monitoring mechanism would allay these concerns. To be credible, this mechanism had to involve some international presence — a highly sensitive issue for the Government of Indonesia, which had to be concerned about attacks from ultra-nationalist politicians.[59] Ahtisaari was able to bring in the EU as a counterweight to ASEAN, providing a level of international involvement in and support for the peace agreement that was critically important to both GAM and the government.

Fourth, Ahtisaari was prepared to risk some of his own political capital when he felt the progress of the negotiations was threatened. "My role was to help make sure the agreement is fair, but also realistic and can endure", he commented, "I have to tell people hard truths." This included being "very tough on GAM about this framework [for the Helsinki negotiations]. I was not afraid to tell them the hard facts: I don't see one single government in the world that supports you".[60] As a result, some on the GAM side were inclined to see Ahtisaari as condescending and favouring the national government in Jakarta.[61]

Ahtisaari, however, was also prepared to challenge Jakarta at the highest levels. Responding to "alarming and very credible reports" about escalating human rights abuses he received from GAM, Ahtisaari travelled to Jakarta in May 2005. "I wanted to bring these directly to the attention of SBY. I myself edited the reports so that the source of the information could not be determined, and then handed them directly to the government. I told the government, 'Get rid of the worst offenders. If you cannot punish them, then at least transfer them'".[62]

THE DECEMBER 2006 LOCAL ELECTIONS

If the success of the Helsinki negotiations was greeted with a mixture of surprise and relief, the prospects for a negotiated peace actually taking hold and weathering challenges from potential spoilers still had to be demonstrated.

Previous peace negotiations had eroded relatively quickly, taking Aceh back into renewed (and often escalated) military conflict. This section examines the period leading up to the successful local government elections in December 2006. It does not provide a detailed chronicle, but instead tries to illuminate how the factors that lead to success in Helsinki were also critical in helping the peace to take hold and endure.

Run-up to the Elections: The Eroding Coherence, Focus and Discipline of GAM

Throughout the post-Helsinki period, GAM's conscious policy was to continue to demonstrate its internal coherence and discipline by clearly meeting its commitments under the MoU. The AMM credited GAM with having demonstrated great discipline in adhering to the terms of the agreement.[63] Although GAM leaders expressed frustration that the government had not met its commitments in a number of important areas (such as assistance for demobilized GAM fighters and key provisions of the recent Law on Governing Aceh), GAM itself carefully kept its objections within the framework of the Helsinki agreement: "We want to be clear that the failure is on the side of the government, which is not meeting its obligations".[64]

In August 2006, I wrote that "there are reasons to question whether this discipline can be sustained for the future. In direct contrast to the national government, GAM now faces a range of new challenges with fewer resources".[65] There were several reasons for this speculation about the difficult future of GAM. With the disarmament and demobilization of GAM forces, it lost its structure of command and control. It was no longer an armed insurgency, and the GAM leadership could no longer rely on a cadre of field commanders and troops. In a very real sense, as one GAM leader observed several months after the MoU was signed, "GAM is finished in the sense that this is an irreversible change. We cannot go back to fighting".[66]

If it was not possible to go back to fighting, the path forward was also not particularly clear. In May 2006, GAM was unable to reach agreement on a common slate of candidates for the local elections. As a result, GAM declared that it would not launch its own political party, nor would it field any candidates formally endorsed by GAM. It thus acknowledged that it had been unable to develop a common position on its future in the post-Helsinki world.

In the subsequent months, a younger generation of GAM leaders emerged as independent candidates, implicitly challenging the established, old guard leaders such as Malik Mahmud.[67] In June 2006, Irwandi Jusuf, a leader of

GAM's intelligence operations during the conflict and a frequent hardline critic of the Jakarta government, launched his independent candidacy for governor. He subsequently received a large plurality of votes in the December elections, with support drawn fairly evenly from across the whole province, and was inaugurated as Aceh's new governor in February 2007.

Successful Local Elections

The December 2006 local government elections were a major milestone for Aceh, marking the transition to a popularly elected government in the new era of peace launched by the Helsinki MoU. Expectations were high. A survey of public attitudes in September–October 2006 indicated that 93 per cent of respondents expressed confidence that the elections would accurately reflect the popular will, with the same percentage believing that they could help secure peace.[68]

The success of the elections was not a foregone conclusion. There were enormous logistical challenges in organizing them. Popular understanding of election procedures was very limited, with the IFES survey indicating that 79 per cent of respondents reported that they did not have very much or any information on the election.[69] In addition, the early informal campaign period was marked by worrisome clashes between competing factions of GAM Bireuen on 27 November. This increased concerns not simply because the elections themselves might be disrupted, but that they could spark further violence.

The IFES opinion poll, however, was also very suggestive of the relative progress of Aceh's journey from combat zone to democratic polity. Most of those who were concerned about violence expected it to come from political parties and independent candidates, while only 14 per cent cited the national security forces and 9 per cent cited GAM as a source of concern.[70] Most notable is the absence of the TNI as a threat to the free expression of the political will of the Acehnese people. In contrast to the April 2004 elections, when there were widespread complaints about TNI control and manipulation of the electoral process and its unwillingness to open the process to international observers, the 2006 elections saw virtually no complaints about TNI misconduct.

In the end, the elections themselves proceeded relatively smoothly, with few reports of intimidation, interference or administrative flaws. About 2.1 million voters, slightly more than 80 per cent of those registered, participated. Reports from the press, international and domestic election observers and NGOs portrayed a generally open environment during the campaign, in which freedoms of expression, association and assembly were respected; a high level of confidence in the organization of the election by the Independent Election Commission (KIP); very few registration irregularities; and almost no reports of intimidation during the voting process or fraud following it.[71]

Acceptance of the Results

If the relatively peaceful campaign and orderly voting were a surprise, the results were even more of a surprise. Virtually all political analysts, academics and the news media expected that no candidate would win sufficient support in the first round to claim outright victory. A second round was widely anticipated.[72] However, the candidates representing the "Young Turks" wing of GAM, Irwandi Jusuf and his running mate, Muhammad Nazar, won nearly 39 per cent of the vote.

This result could have raised considerable anxiety in Jakarta. Irwandi was widely seen as a GAM hardliner who had reluctantly acquiesced to the fundamental premise of the Helsinki MoU, that Aceh would remain an integral part of the unitary republic. He had been critical of the provisions of the 2006 Basic Law on the Governance of Aceh (Undang Undang Pemerintahan Aceh, or UUPA) and many in Jakarta were inclined to speculate that he had not really relinquished the aspiration for eventual complete independence. His running mate, Muhammad Nazar, was seen by some as even more of a hardliner. Prior to the Helsinki MoU, both had been imprisoned by Jakarta, with Irwandi managing a dramatic escape when the prison in which he was held was destroyed by the tsunami.

With this background, it would not be surprising for the political elite in Jakarta to view the Irwandi/Nazar victory with some alarm.[73] However, SBY and Kalla moved quickly to express clear and uniform respect for the results and a pragmatic willingness to work with whomever had been selected by the Acehnese people as their new governor. What was important for the president was that the elections ran smoothly and transparently. International support reinforced respect for the results.[74] Minister of Defence Juwono Sudarsono even went so far as to strike an apologetic note, stating that the election results are "an expression of a desire for autonomy by the Acehnese. The message is that the central government must be more attentive to the Aceh people … The people in Aceh are, as it turns out, not satisfied with how they were treated in the past. We accept this, and the central government needs to be prudent in developing Aceh".[75]

These statements were met by reassurances from Irwandi Jusuf about his intentions: "Jakarta has nothing to worry about. Everything has been regulated in the memorandum of understanding (MoU) that Indonesia and GAM signed in Helsinki on 15 August 2005. I do not like political rhetoric, and government officials in Java should not use rhetoric and make unnecessary comments on such issues. GAM and all its supporters are strongly committed to the peace accord in order to build a permanent peace and improve the lives of the Acehnese, as an integral part of Indonesia".[76] Irwandi has continued to reach out to earlier opponents, most recently meeting with army chief of staff General Djoko Santoso to seek support in leading the province. According to press reports, Irwandi stressed that support from the military, especially the army, would be indispensable in helping Aceh's recovery.[77]

In the period ahead, it is an open question whether it is useful or accurate to describe the newly elected governor of Aceh as a GAM leader, or if he will create a new political organization. Similarly, it is unclear if GAM can or will seek to develop its own political programme and structure, or if it will limit itself to advocating for the interests of its former troops in the reintegration assistance programmes.

BEYOND HELSINKI: LESSONS AND PROSPECTS FOR THE FUTURE

A Stepping Stone to Improved National Governance?

The ability of the SBY/Kalla administration to bring greater coherence, focus and discipline to the government's approach to Aceh has given them increased credibility and authority within the broader context of Indonesia's democratic reforms. SBY and Kalla have demonstrated that they are able to enforce greater discipline and impose significant sanctions on those who try to undermine government policies. This also helps explain why the government has (largely) met its commitments under the MoU, and avoided the subversion and sabotage that characterized earlier peace efforts.

The significance of SBY's decision to recall the nomination of Ryacudu and his subsequent steps to bring the TNI on board thus extends beyond the Helsinki negotiations. By denying Ryacudu the supreme command position in the TNI in the first few weeks of his administration, and later replacing him as army chief of staff in February 2005, SBY was asserting his leadership and control as the civilian president over the military: "Yudhoyono's success in enforcing military compliance in Aceh marked a watershed in post-Suharto civil-military relations. For the first time, the government was able to secure the military's support for a negotiated settlement with separatist rebels".[78]

Yudhoyono's choice of successor to Endriartono Sutarto, Air Chief Marshal Djoko Suyanto, in early 2006 confirmed his policy of continuing the gradual reduction of the involvement of the TNI in political affairs. During his confirmation hearings before the DPR, Marshal Suyanto vowed to stay out of politics and to protect human rights.[79] Djoko Suyanto was the first air force chief of staff to be nominated to lead the military, a position that has traditionally been held by the army. Although Megawati's PDI-P continued to criticize Yudhoyono's decision to bypass Ryacudu, Djoko still received overwhelming support from the DPR.[80]

More recently, Djoko has unveiled a new set of guidelines as a part of his pledge to the DPR to bring the TNI into conformity with the 2004 military reform law. In January 2007, he announced a new doctrine that bans the armed forces from active involvement in the country's sociopolitical affairs. As quoted in the *Jakarta Post*, Djoko stated that "in the past, military personnel could get involved in politics because the old doctrine allowed them to do so. Now we no longer associate ourselves with politics".[81]

Significant military reforms remain to be implemented, most notably the overhaul of the non-transparent military financing by companies and foundations owned by the TNI and reform of territorial structure. It is possible that SBY has relied more on personal connections than on institutional reforms, and that the process of reform is incomplete and "counterbalanced by serious omissions and failures".[82] Nonetheless, the resolution of the conflict in Aceh established his commitment to broader civilian control and strengthening of democratic governance in the nation as a whole. The question for the future is how SBY and Kalla will build on this achievement. They appear to be poised for the next step forward, but it is unclear that they have yet exploited this advantage and applied it to other key areas.

The Promise and Perils of Ad Hoc Approaches

As noted earlier, part of the reason for the success of the Helsinki negotiations was the decision on the part of SBY and Kalla to work through a small ad hoc group, keeping discussions directly under their oversight and out of any formal channels of the bureaucracy. However, this is not necessarily a recipe for long-term success. The transition from relatively intense but brief negotiations in Helsinki to the long-term and more prosaic challenges of administrating large, complex and ongoing programmes probably requires a different kind of response. Ad hoc approaches that bypass the bureaucracy and rely on highly personal, informal networks are likely to become a liability rather than an asset.

During the period from the Helsinki MoU in August 2005 to the early election campaign in September 2006, it was not always clear who was responsible for fulfilling various requirements of the MoU. The AMM tended to emphasize security issues, through its Committee on Security Affairs, with representatives of the TNI and GAM. Over time, however, other issues began to emerge. These included ensuring that sufficient funding was available for supporting the re-integration of ex-combatants, and that the ground rules for offering compensation to victims of the conflict were clear and consistently implemented. Here, the transition from an ad hoc negotiating team to the established bureaucracy was sometimes difficult. By the March–May 2006 period, the release of funds for re-integration programmes were caught up in bureaucratic delays, and GAM leaders began to fret that the obligations of the national government were not being fulfilled.[83]

Although very large sums had been earmarked to support re-integration programmes, it was not always easy to identify any specific agency and then to hold it accountable for implementation. There seemed to be a proliferation of bodies that could claim some role in this process, each with access to different sources of funds, managed by different groups, operating with different mandates and articulating different priorities and procedures. The Aceh Rehabilitation and Reconstruction Agency (Badan Rehabilitasi dan Rekonstruksi Aceh, or BRR) was initially established to oversee and coordinate the substantial international assistance after the December 2004 tsunami, but the targets of this assistance included both geographic areas affected by the conflict in Aceh and individuals who could claim to be victims of both the conflict and the tsunami. The acting governor of Aceh then established the Aceh Reintegration Body (Badan Reintegrasi Aceh, or BRA), which was intended to coordinate and manage assistance and compensation directly related to the implementation of the Helsinki MoU. The Joint Peace Forum (Forum Bersama Damaia, or Forbes Damai) included the government, GAM and international donors, and also was to coordinate assistance. All of these were ad hoc bodies, specially created to bypass the established systems of public administration and ensure the smooth implementation of reconstruction programmes.

At the national level, the cohesion of the negotiating team began to be diffused. With the conclusion of the Helsinki negotiations, the role of Hamid Awaluddin appeared to diminish, while Sofyan Djalal seemed to emerge as the key spokesman and interlocutor for the government. As minister of Information, he had no direct control over the administration of funds through BAPPENAS and the provincial government. Djalal was active in meeting, nudging and cajoling, but he could not exercise any direct authority. The challenge for incoming Governor Irwandi is to assert some control over these diverse bodies, and move their ad hoc planning and decision-making into a single process of governance that makes the aspiration of significant autonomy for Aceh a reality.

Aceh and Poso?

In January 2007, conflict in the Central Sulawesi city of Poso erupted, with battles between the police and a group of local jihadi suspects. One policeman and fifteen others were killed, including some ordinary residents who seem to have been caught in the wrong place at the wrong time. As well as a traumatic event for the local community, this had to be a great personal disappointment to Jusuf Kalla. It was the widely perceived success of his earlier efforts to resolve conflict in this region in 2001 that gave him both confidence and credibility in tackling Aceh. Now, that early success seems to be threatening to unravel at virtually the same time that Aceh is marking another step forward in its peaceful evolution.

Yet, the recent detailed account of these events by the International Crisis Group (ICG) points to both an important similarity with as well as a significant divergence from the Aceh peace process as examined in this chapter. The similarity is the concern over possible involvement of the TNI in the conflict, and its potential to exacerbate rather than defuse the problem. The significant difference is that the 2001 Molino Accords had no equivalent of the AMM to monitor implementation of the agreement, investigate alleged violations, goad the conflicting parties into fulfilling their obligations, and hold accountable those within their own ranks who undermined the agreement. These lessons from the Aceh peace process might contribute to a durable resolution of the Poso conflict if they were incorporated into the government's approach.

Aceh Is Not Papua

Following the success of the Helsinki negotiations, there has been a natural tendency of many observers to turn to Indonesia's other major separatist conflict in Papua. If a lasting settlement is possible in Aceh, why not in Papua? Like Aceh, Papua has experienced a long-standing, restive and often violent conflict between pro-independence groups and the national government. Like Aceh, the appeal of pro-independence forces (as exemplified by the OPM, the Free Papua Organization, or Organisasi Papua Merdeka) has been to a separate national and cultural identity which is historically distinct from the mainstream of Indonesian national identity.[84]

But there are very significant differences that suggest that the path to success in Aceh may not be relevant to Papua. First, OPM is not GAM. It does not have the same coherence, discipline and broad (although not unchallenged) claim to represent the Papuan people. As was noted earlier, the coherence and discipline of GAM was a significant factor in the ultimate success of the Helsinki negotiations. It is hard to find the equivalent of GAM in Papua. OPM does not control territory to the same extent as GAM. Around 2000–04, GAM was running a parallel government in many areas, with its own system of taxation and law enforcement. This was greatly reduced under the pressure of military operations following the 2004 declaration of martial law, but neither OPM nor any other group has established this kind of territorial control.

For better or worse, the lack of significant military success or territorial control by the OPM has meant that there has been no clear negotiating partner for the national government. If the government were willing to enter into negotiations over the status of Papua, with whom would they negotiate? Only with the relatively recent creation of the Papuan People's Assembly (Majalis Rakyat Papua, or MPR) has there been a single recognised forum for the expression of Papuan views. The limited experience thus far suggests that within the Papuan political elite there is nothing like the shared vision that helped hold GAM together over several decades.

CONCLUSION

This chapter has argued that the key factors that enabled the negotiators in Helsinki to reach agreement laid the foundation for the success of the December 2006 local government elections. The strong conviction of SBY and Kalla, that no military solution was possible and that some political settlement was inevitable, was the critical starting point. Based on this conviction, SBY was prepared to make difficult decisions to reduce the scope of the TNI for undermining negotiations or threatening the viability of the peace agreement. In turn, this gave Kalla the political space necessary for assembling a small team of trusted confidants to conduct negotiations, working outside official channels. The close collaboration with SBY and Kalla enabled the government to bring policy coherence and discipline to the government side that was lacking in earlier attempts to find a peaceful settlement.

This new coherence on the part of the government was matched by the extraordinary discipline and pragmatism of the GAM leadership. After considerable internal debate, they made the difficult decision to relinquish their aspirations for complete independence, but at the same time pushed hard to expand the scope of Acehnese control over their own lives, within the framework of the Indonesian state but without the looming interference of Jakarta. The GAM leadership was also able to carry their field commanders and political cadres with them into this new agreement and to ensure their adherence to the requirements of the MoU.

Both sides were assisted by the almost accidental engagement of international facilitators, who helped established communication, provided the structure for negotiations, and ensured credible international support and mechanisms for monitoring implementation. The tsunami, with both its appalling human costs and enormous international assistance, was a strong motivating factor, but the groundwork for the Helsinki process started long before the tsunami, and the ability to reach an agreement was more directly the result of deliberate choices and policies that were already in place before it struck.

The key factors behind the success in Helsinki not only gave both parties confidence and credibility, but also helped them stay the course through the ensuing months. Thus, the December 2006 elections, marking a new phase in Aceh's continuing path towards peace, was a direct result of the same factors that made for the initial success of negotiations in Helsinki during the January–August 2005 period.

The Helsinki MoU was important not only for Aceh but also for Indonesia's broader democratic development. This is particularly true with regard to increased civil control over the military and continuing efforts to reform the military. In this sense, all Indonesians, and all of Indonesia's many friends in the international community, have good reason to celebrate both Helsinki and the journey that stretches beyond it.

Notes

1. An earlier version of this paper, entitled "Beyond Helsinki: Aceh and Indonesia's Democratic Development", was prepared for the First International Conference on Aceh and Indian Ocean Studies held in Banda Aceh, Indonesia, 24–26 February 2007.
2. See especially Edward Aspinall, *The Helsinki Agreement: A More Promising Basis for Peace in Aceh?* Policy Studies 20 (Washington, DC: East-West Center, 2005); Edward Aspinall and Harold Crouch, *The Aceh Peace Process: Why it Failed*, Policy Studies 1 (Washington, DC: East-West Center, 2003).
3. Damien Kingsbury, *Peace in Aceh: A Personal Account of the Helsinki Peace Process* (Jakarta: Equinox Publishing, 2006) presents a highly personal account from the perspective of a GAM adviser, and thus has relatively little to say about

the dynamics on the side of the Indonesian Government. Finnish journalist Katri Marikallio is about to publish an account of the negotiations, drawing on interviews with both sides, and so is Finnish mediator and former president Martii Ahtisaari (*Making Peace: Aceh and Ahtisaari*, forthcoming).

4. International Crisis Group (ICG), "Aceh's Local Elections: The Role of the Free Aceh Movement (GAM)", ICG Asia Briefing No. 57 (Jakarta/Brussels: ICG, 2006).

5. See, for example, the article by Sandra Hamid and Douglas Ramage, "Autonomy for Aceh", *Wall Street Journal*, 18 July 2006, which cites the tsunami as the key factor "that finally persuaded both sides to put an end to the conflict".

6. Aspinall, *Helsinki Agreement*, pp. 7–10; International Crisis Group (ICG), "Aceh: A New Chance for Peace", ICG Asia Briefing No. 40 (Jakarta/Brussels: ICG, 2005) also cites the impact of military operations on GAM.

7. Reports from the ICG, for example, prominently feature the role of Jusuf Kalla, referring to "the Kalla initiative". See ICG, "Aceh: A New Chance for Peace", pp. 1–4. Similarly, Aspinall describes Kalla as "the most active government advocate of the talks". See Aspinall, *Helsinki Agreement*, p. 14.

8. Separate author interviews with Juha Christensen (16 May 2005) and Martii Ahtisaari (21 June 2005) have confirmed Aspinall's chronology. See especially Aspinall, *Helsinki Agreement*, p. 19.

9. This certainly was the assessment of the government, which made a clear effort to undercut the security of the GAM leadership in Sweden by initiating efforts to have them expelled (author interview with Sofyan Djalil, 13 May 2006, Jakarta, Indonesia).

10. Damien Kingsbury's book perhaps makes this point most strongly. Although a strong advocate for GAM and occupying a unique position as a foreign adviser actively engaged throughout the Helsinki negotiations, his impatience, frustration and even despair over what he saw as the sloppy preparations and thinking on the GAM side are a recurring theme. See Kingsbury, *Peace in Aceh*.

11. Aspinall, *Helsinki Agreement*, pp. 15–19, gives a good overview of many of these efforts, although author interviews with Juha Christensen (16 and 17 May 2005) suggest some differences in detail. However, the main point, that SBY and Kalla were very interested in re-establishing a dialogue with GAM and were actively seeking ways of achieving this, is not in doubt.

12. Author interview with Jusuf Kalla, 25 July 2005, Jakarta, Indonesia.

13. Author interview with Farid Husain, 23 May 2005, Jakarta, Indonesia.

14. Author interview with Farid Husain, 23 May 2005, Jakarta, Indonesia.

15. Author interview with Pieter Feith, 19 July 2006, Banda Aceh, Indonesia.

16. Author interview with Farid Husain, 23 May 2006, Jakarta, Indonesia.

17. Author interview with Susilo Bambang Yudhoyono, 21 May 2006, Jakarta, Indonesia.

18. Aspinall and Crouch, *Aceh Peace Process*, pp. 23–24.

19. Hamid and Ramage, "Autonomy for Aceh".

20. This suggests that unqualified enthusiasm for strengthened democratic processes may be too uncritical or even naive. Contrary to arguments such as those advanced by Hamid and Ramage in "Autonomy for Aceh", the active and vociferous national political parties in the DPR constituted a challenge to be overcome in reaching agreement in Helsinki, rather than a fundamental cause of success of the Helsinki process.

21. See, for example, Aspinall and Crouch, *Aceh Peace Process*, pp. 29–30.

22. Author interview with Susilo Bambang Yudhoyono, 23 May 2006, Jakarta, Indonesia.

23. "Perdamaian di Aceh Saatnya Diwujudkan", *Kompas,* 13 June 2005. Cited in Marcus Mietzner, *The Politics of Military Reform in Post-Suharto Indonesia: Elite Conflict, Nationalism, and Institutional Resistance,* Policy Studies no. 23 (Washington, DC: East-West Center, 2006) <http://www.eastwestcenter.org/fileadmin/stored/pdfs/PS023.pdf> (accessed 19 May 2009).

24. "TNI Sanggup Selesaikan Masalah Aceh", *Kompas,* 5 April 2003.

25. Aspinall and Crouch, *Aceh Peace Process*, p. 54.

26. Author interview with Susilo Bambang Yudhoyono, 21 May 2006, Jakarta, Indonesia.

27. This is certainly one of the main themes emerging from the analysis of Aspinall and Crouch. See Aspinall and Crouch, *Aceh Peace Process*, pp. 2–5, 16–17, 23–24, 29–30, 40–42.

28. Author interview with Susilo Bambang Yudhoyono, 21 May 2006, Jakarta, Indonesia.

29. Ibid.

30. Of all the interviews conducted for this study, Kalla alone produced thick volumes of files, each carefully organized and tabulated, and filled with his memos, marginal comments, revisions and recommendations. He stated that he had drafted most of the documents himself with little assistance from staff.

31. Author interview with Susilo Bambang Yudhoyono, 21 May 2006, Jakarta, Indonesia.

32. Author interview with Jusuf Kalla, 25 July 2006, Jakarta, Indonesia.

33. For example, see ICG, "Aceh: Escalating Tension", ICG Indonesia Briefing No. 4 (Banda Aceh/Jakarta/Brussels: International Crisis Group, 2000), pp. 4, 7–8; ICG, "Aceh: Why Military Force Won't Bring Lasting Peace", ICG Asia Report No. 17 (Jakarta/Brussels: International Crisis Group, 2001), pp. 12–15; Aspinall and Crouch, *The Aceh Peace Process*, p. 35.

34. Shulze, "Insurgency and Counter-Insurgency", quoting from *Kompas,* 20 May 2003.

35. Damien Kingsbury and Lesley McCulloch, "Military Business in Aceh", in *Verandah of Violence: The Background to the Aceh Problem,* edited by Anthony Reid (Singapore: Singapore University Press, 2006).

36. Mietzner, *The Politics of Military Reform,* pp. 49–50.

37. Author interviews with Juwono Sudarsono, 23 May 2006, and Sofyan Djalil, 13 May 2006, Jakarta, Indonesia.
38. Author interview with Susilo Bambang Yudhoyono, 21 May 2006, Jakarta, Indonesia.
39. Author interview with Sofyan Djalil, 13 May 2006, and Juwono Sudarsono, 23 May 2006, Jakarta, Indonesia.
40. Author interview with Susilo Bambang Yudhoyono, 23 May 2006, Jakarta, Indonesia.
41. Ibid.
42. Ibid.
43. Author interview with Susilo Bambang Yudhoyono, 21 and 23 May 2006, Jakarta, Indonesia.
44. Author interviews with Susilo Bambang Yudhoyono, 21 May 2006, Jakarta, Indonesia.
45. Ibid.
46. Ibid.
47. Author interview with Jusuf Kalla, 25 July 2006, Jakarta, Indonesia.
48. Aspinall and Crouch, *Aceh Peace Process*, p. 23.
49. Author interview with Sofyan Djalil, 13 May 2006, Jakarta, Indonesia. Ironically, Djalil says that his own Acehnese origins may have been viewed negatively by some on the GAM delegation whom he thinks might have regarded him as a traitor to his people. Nonetheless, as negotiations proceeded, he was able to revert to informal conversation in Acehnese, and also to help the government delegation interpret some of the GAM approaches, positions and sentiments.
50. Ibid.
51. Author interview with Widodo Adi Sudjipto, 24 May 2006, Jakarta, Indonesia. Interestingly, Widodo referred to his role in the delegation as "supervisor" acting on behalf of the president, suggesting he was there to make sure that "Kalla's men" did not wander from the agreed framework. However, others — including Kalla himself — have suggested that Widodo was uncomfortable that Hamid (with a more junior rank) was leading the delegation.
52. I am using the term "amateur" here to indicate individuals who were not professionally trained "peacemakers" associated with institutions whose mission was mediating disputes. They did not come to the Helsinki process with a canon of established knowledge about brokering disputes or a tested toolkit that had been effective in other disputes. They were highly committed, energetic and skilled in their own ways, but they were not credentialled mediators.
53. Author interviews with Susilo Bambang Yudhoyono, 21 May 2006; Sofyan Djalil, 13 May 2006; and Malik Mahmud, 17 May 2006, Jakarta, Indonesia.
54. Author interview with Malik Mahmud, 17 May 2006, Banda Aceh, Indonesia.
55. Author interview with Martii Ahtisaari, 21 May 2006, Helsinki.

56. Kingsbury, *Peace in Aceh*, p. 26. Also author interview with Martii Ahtisaari, 21 May 2006, Helsinki.
57. Author interview with Martii Ahtisaari, 21 May 2006, Helsinki.
58. Author interview with Martii Ahtisaari, 21 May 2006, and Meeri-Maria Jaarva, 21 June 2006, Helsinki.
59. Author interview with Susilo Bambang Yudhoyono, 21 May 2006, Jakarta, Indonesia.
60. Author interview with Martii Ahtisaari, 21 May 2006, Helsinki.
61. Kingsbury, *Peace in Aceh*, p. 16. Also author interview with Nur Djuli, 17 July 2006, Banda Aceh, Indonesia.
62. Author interview with Martii Ahtisaari, 21 May 2006, Helsinki.
63. Author interview with Pieter Feith, 19 July 2006, Banda Aceh, Indonesia.
64. Author interview with Irwandi Jusuf, 19 July 2006, Banda Aceh, Indonesia.
65. Michael Morfit, "Staying on the Road to Helsinki: Why the Aceh Agreement was Possible in August 2005", paper prepared for the international conference on "Building Permanent Peace in Aceh: One Year After the Helsinki Accord", sponsored by the Indonesian Council for World Affairs (ICWA), Jakarta, Indonesia, 14 August 2006, p. 25.
66. Author interview with Nur Djuli, 17 July 2006, Banda Aceh, Indonesia.
67. ICG, "Aceh's Local Elections", pp. 3–6.
68. IFES, *Opinions and Information on the Pilkada Aceh Elections 2006: Key Findings from an IFES Survey*, 1 December 2006, p. 4 <http://www.ifes.org/publication/da48709050aa294dfdc0cda43969420b/IFES%20Aceh%20Pilkada%20Survey%201%20Exec%20Summary.pdf> (accessed 19 May 2009).
69. IFES, "Opinions and Information", pp. 2–3.
70. IFES, "Opinions and Information", p. 4.
71. European Union Election Observation Mission, "Statement of Preliminary Conclusions and Findings", 12 December 2006, Aceh <http://www.eueomaceh.org/Files/ACEH_Preliminary_Statement_English.pdf> (accessed 19 May 2009).
72. For example, see the report by Achmad Sukarsono, "Landmark Aceh Election Expected to Face Run-Offs", Reuters UK, 6 December 2005 <http://www.alertnet.org/thenews/newsdesk/JAK276134.htm> (accessed 19 May 2009).
73. See, for example, "Ex-Rebel Wins Aceh Election", *ABC News Online*, 12 December 2006 <http://www.abc.net.au/news/newsitems/200612/s1809307.htm> (accessed 19 May 2009).
74. See, for example, the statement by UN Secretary General Kofi Annan "calling on all sides to respect the landmark election results", United Nations Press Release, 11 December 2006 <http://www.un.org/News/Press/docs/2006/sgsm10791.doc.htm> (accessed 19 May 2009).
75. "Aceh Poll Underlines Popularity of GAM", *Jakarta Post*, 14 December 2006.
76. "Aceh's Future Governor Plays Down Separatism Jitters", *Jakarta Post*, 19 December 2006.

77. "Irwandi Seeks Army Support", *Jakarta Post*, 6 February. 2007.
78. Mietzner, *Politics of Military Reform*, p. 51.
79. "Indonesian Military's New Day?" *International Herald Tribune*, 2 February 2006. See also "Yudhoyono's Choice of Military Chief Approved", *Straits Times*, 3 February 2006.
80. "Military Must Have a Presence in the Region", *Jakarta Post*, 3 February 2006.
81. "TNI Unveils New Doctrine: No Politics", *Jakarta Post*, 25 January 2007.
82. Mietzner, *Politics of Military Reform*, pp. viii, 59–66.
83. Author interview with Irwandi Jusuf, 19 July 2006, Banda Aceh, Indonesia.
84. Blair King, "Peace in Papua: Widening a Window of Opportunity", Council on Foreign Relations, Center for Preventative Action No. 14 (New York: Council on Foreign Relations, 2006), pp. 6–7.

10

MAKING PEACE AGREEMENTS EFFECTIVE
The Aceh Monitoring Mission Experience

Pieter Feith

INTRODUCTION

The people of Aceh have been the victims of two immense disasters: more than three decades of conflict that officially ended in 2006 and the 2004 tsunami. Both of these have been at the centre of immense international focus since the tsunami propelled Aceh onto the world stage; the peace process and the post-tsunami relief and reconstruction efforts have involved significant amounts of international intervention. Both are important examples that need to be better understood, as they can provide valuable lessons for other conflict and post-disaster situations in the region and beyond.

In this chapter, I provide a brief assessment of the workings of the peace process from the perspective of the Aceh Monitoring Mission (AMM), which I had the honour of heading from 2005 until 2006.[1] The AMM was put in place as part of efforts to ensure that both GAM and the Indonesian Government operated in good faith on the obligations agreed to in the Helsinki Memorandum of Understanding, signed on 15 August 2005. I explore the experiences of the AMM and discuss why the peace process has so far been relatively successful. Following this, I will talk about the wider implications this process has for Indonesia and the region, and about the importance of peace in the reconstruction and development of Aceh. Finally, I conclude by discussing some matters that might be factors in the future.

POLITICAL FACTORS FOR THE PEACEFUL SOLUTION: FOUR KEY REASONS

While the cessation of hostilities in Aceh, like all peace processes, was an incredibly complicated process that required the sincere efforts of multiple parties to resolve, there are four key reasons for its success that I want emphasize. These points are crucial elements without which the peace process would not have succeeded.

First, the MoU signed in Helsinki was very well constructed. Not only did it address the main concerns of both parties, but it also benefited from clearly articulated sets of provisions, responsibilities, and timelines. It was a document with real substance and reasonable steps that could be undertaken by both parties. These steps were measurable by the inclusion of reliable benchmarks, so that mutual progress could easily be monitored in a way that was transparent to the parties directly involved (the Acehnese, GAM, the Indonesian Government, and the Indonesian military) and to the wider international community. Such provisions created a significant synergy between the negotiators and the subsequent monitors called in to oversee the implementation of the process. This is critical in such complex processes, as there is often much room for miscommunication that can threaten the whole process. Bringing the parties to agreement, both on the terms of the accord and how it was going to be implemented, was a major and necessary task.

Secondly, both parties exerted strong political will to make the negotiations and the implementation work. This was facilitated by each being flexible on the major points that, if not resolved, would have terminated the talks. GAM's main concession was giving up its previous end-game of political independence. This was a clear deal-breaker for the Indonesian Government and would have been an insurmountable obstacle to the negotiations. This was a major concession from GAM, as it effectively conceded that Aceh is an integral part of Indonesia, and ran counter to much of their rhetoric in the previous three decades of struggle for independence. Furthermore, for a separatist movement dispersed in exile, GAM showed remarkable discipline and cohesiveness as an organization, with strong central leadership. This enabled them to get support on the ground and amongst enough of their supporters to make such bold concessions and still maintain influence. This was a major reason for the success of the talks.

On the government side, the Indonesian president, Susilo Bambang Yudhoyono (SBY), and Vice President Jusuf Kalla were deeply committed to the success of the process and undertook significant steps to make it work, as discussed in great detail by Morfit in this volume. This involved taking a firm stance on hardliners, expending political capital on a potentially risky and unpopular venture and, perhaps most importantly, working to reform the Indonesian military. These steps were necessary for the devolution of power and to make the issue firmly a political one that required a negotiated political solution rather than a military one. SBY accepted that there was most likely not an imminent military solution to the situation with Aceh and that it was essential to try other routes to end the conflict. These steps were just as significant, in terms of the change that they represented, as the main concessions that GAM made — which also provided balance in terms of what each side had to offer. Peace in such situations usually requires each party giving far more than they would like; in some regards, both GAM and the Indonesian Government opted to do so.

Third, the monitoring was led by a coalition of states; it was not left to the oversight of an NGO or consortium of respected individuals. The states involved proved both to be credible and also had suitable leverage over the parties to keep both sides in line and on track. In particular, the framework of the agreement allowed the AMM to constantly gauge the progress of the process and facilitate problem-solving. This unique arrangement gave the process credibility, international weight and also operational flexibility to oversee and deal with issues as they arose on the ground. Additionally, there are a range of resources that state actors, and especially collections of states such as the EU, are able to muster that are necessary for mobilizing different forms of support, such as broader packages of economic development aid and dealing with potentially isolated issues within a much broader international political framework.

Finally, it is clear that the majority of the people in Aceh wanted peace. Many of the Acehnese were tired of endless conflict and put pressure on the parties to find a solution with dignity, so that they could build their lives and livelihoods. Just as SBY and Kalla acknowledged that there was not a viable military solution to the troubles in Aceh, the majority of the Acehnese also came to the same realization. It had become clear that armed struggle could not achieve a permanent solution of Acehnese independence and would most likely just keep Aceh in a perpetual state of low-intensity conflict defined by isolation and underdevelopment. This was compounded by the destruction of the tsunami, which brought a whole new set of dynamics into the mix, as discussed in more detail by other contributors to this volume.

AMM — INNOVATIVE CRISIS MANAGEMENT

The AMM was the first European Security and Defence Policy (ESDP) mission in Asia. The basic aim of the ESDP is to:

> allow the [European] Union to develop its civilian and military capacities for crisis management and conflict prevention on an international level, thus helping to maintain peace and international security, in accordance with the United Nations Charter. The ESDP, which does not involve the creation of a European army, is developing in a manner that is compatible and coordinated with NATO.[2]

The efforts in Aceh are part of the EU's broader policy context of strengthening security and stability in Southeast Asia. This has been achieved partly by a new level of cooperation between the EU and ASEAN to build lasting partnerships, aimed at jointly managing both regional threats and situations that are, or could become, global problems. It is becoming an essential part of the EU mission to look outside of Europe and to participate in strategic partnerships to encourage multilateralism, increase security and facilitate development. This fits into the broader EU goal of becoming a meaningful global player, without posturing to be a superpower. The EU hopes that through its political and economic weight, resources and shared values — what some would call "soft power" — it can contribute to a safer and more prosperous world.

The core components of the EU's approach are preventing "failed" states and cooperating with regional authorities to combat international terrorism, transnational crime, proliferation of weapons of mass destruction and violent religious extremism. In many cases, these problems are inherently linked, so it is necessary to view them in a more holistic manner and look for comprehensive solutions that target the conditions that breed such issues. Furthermore, it is no longer possible to view problems of this nature as being localized. Increasing mobility and global interconnectedness means that problems can easily transcend the boundaries of territories and nation-states. Fortunately, so too can the solutions.

One of the core strengths of the EU, which can be seen in the make-up of the AMM, is the unique mix of competencies it can draw upon. It is able to tap into both civilian and military expertise. This is especially important, as crisis management in situations such as Aceh requires a broad range of instruments to enact solutions. This can be seen in how the EU has been involved in Aceh — a complicated mission far from its normal geographic sphere of interest. The European Commission (EC) has backed the Aceh peace process through various means and programmes. It is clear from this experience that the EU has the capabilities to play a significant role in crisis management situations outside of Europe.

CORNERSTONES OF SUCCESS

While the peace process was a very difficult and trying one for all involved, it is worth reflecting on a few of the key points that I feel contributed to its success and that should be considered in future peace talks. I have already mentioned the tools and the crucial political influence that the EU was able to contribute. There were also some major operational factors that allowed for a successful involvement in Aceh. I will discuss these in turn below.

Rapid Deployment

One of the main strengths of the EU efforts in Aceh was the pace of deployment. Interventions in conflict situations are notoriously slow, often constantly in the process of catching up to the ongoing action on the ground. This leads to all sorts of problems. In Aceh, the Initial Monitoring Presence (IMP) was sent immediately to begin work. The IMP's presence on the ground filled a potentially harmful vacuum that might have occurred before the AMM could officially be deployed. Extended gaps between the signing of accords and the presence of a neutral authority to monitor implementation are opportunities for momentum to be lost and parties to renege on agreements. In general, the longer the gap, the higher the potential for things to go wrong in the first delicate stages where there is limited or no trust between the principal parties.

The presence of the IMP filled this gap and ensured that there was constant oversight; it was also a clear reminder that the actions of both parties were being closely watched. Furthermore, by laying down the groundwork, the IMP made it possible for the AMM to start decommissioning GAM's weapons on the first day they arrived in Aceh. This was a major step for GAM and, in many ways, a point of no return for them. Making sure that it happened in a timely manner did much to cement the peace process and make clear to all involved that it was serious break from past patterns of failed discussions. In any conflict resolution situation, it is essential to have rapid action; the IMP and the AMM were able to support this. The practice of immediate presence and early action should become the standard for all future EU missions and a point worth noting for other peace processes.

EU-ASEAN Cooperation

A second important reason for the success in Aceh was that the AMM had fully integrated teams of EU and ASEAN monitors. This allowed both groups to contribute relevant skills and expertise that greatly strengthened the mission. ASEAN participation gave regional legitimacy to the mission, allowing the nations in Southeast Asia to be more firmly involved in the process and committed to its success. Additionally, the Asian monitors had a better sensitivity to and respect for local conditions, circumstances and cultures in Indonesia and Aceh. Many of the Asian monitors were able to speak local languages and were also Muslim. This allowed for a much higher level of meaningful interaction with the local parties, and also for a more comfortable operating environment. The EU monitors were able to benefit tremendously from their involvement. The EU provided much of the macro-level planning and financial framework for the operation. This tapped into their recent experiences dealing with the conflicts, in particular in the Balkans. The EU was also able to provide necessary distance and objectivity, as one would be hard pressed to find a potential conflict of interest in the EU's role in Aceh, whereas regional politics always has the potential to taint local involvement.

I think the AMM should be seen as a possible model for future cooperation between regional players in crisis management. This would be helpful for resolving other conflicts in the Indian Ocean region and Southeast Asia. Furthermore, this model can be employed in many parts of the world, bringing together local parties with external partners to create the best possible conditions for resolving conflict. This not only brings together a good mix of talent, experiences and resources, but also makes it clear to all that there is a definite international interest in the peace processes, which in itself adds weight to the efforts.

Proactive Monitoring

One of the key aspects of the actual monitoring was the emphasis upon proactive monitoring. This was a break from the typical monitoring stance in which the overseeing party plays a more passive role, effectively waiting for something to happen and then to reacting to it. Given our mandate and the nature of the agreements reached, we were able to play more of a leadership role, which allowed us to address potential problems with the two parties before they escalated and became major issues that could derail the peace process. The AMM had the authority to influence the pace of the implementation, to set priorities and to suggest the agenda to the parties through the Commission on Security Arrangements (COSA) framework. This flexibility and authority allowed us to be far more effective at nudging the process along and reacting in a sensible and timely manner to the potential difficulties that are common in all peace processes.

Furthermore, the mission members were constantly in contact with the parties and civil society representatives, proposing new ideas and running a substantial public information campaign. This contributed to making the public feel that they had a role in the process and that their concerns were taken into account. This made the entire process very responsive to the concerns of all the main stakeholders and put the AMM in a position where it was engaged in a meaningful way with all parties.

IMPLICATIONS FOR INDONESIA AND THE REGION

The peace agreement in Aceh was a major change in the political landscape of Indonesia. In many ways, Aceh can be seen as a pilot project for the country. Indonesia — with its thousands of islands, varied religious directions and many different ethnic groups — has always had a cause for potential regional discontent. It has long been plagued with different forms of tension, social unrest and larger-scale questions about autonomy and independence from different regions, religious and ethnic groups. Coming to terms with the "Aceh problem" has allowed the Republic of Indonesia to experiment with different approaches, and its success could provide both blueprints and momentum for addressing other regional issues. "Federalism" is no longer a taboo. The Law on the Governing of Aceh (LoGA) was seen in Jakarta as a possible model for autonomy legislation for other parts of Indonesia. Furthermore, the new approach of Jakarta, including the impressive reforms of the military and the improvement of governance, can potentially move Indonesia further down the road to democratic development and stability.

A number of key changes should be discussed. The first and perhaps most important reform has been the increasing subordination of the TNI (Tentara Nasional Indonesia) to elected civilian authority. The role of TNI has changed, and a clear division of labour between the military and police has come into effect, even if, in some places, implementation needs more time. In large part, this has been possible because there was significant recognition within the ranks of the military of the need for reform and more accountability to civilian authority. The Government of Indonesia took unprecedented steps when it publically agreed to joint COSA statements condemning the TNI for excesses committed in Aceh. Also, a number of key military leadership shared SBY's belief that there was no feasible military solution to the problem in Aceh (granted, there are also others in the TNI who are opposed to such reforms and hold a much harder line on responses to separatist movements in Indonesia).

As discussed Morfit in this volume, SBY and Kalla personally assured the best possible conditions for the mission to bring the process forward and towards a successful conclusion. AMM was able to report on Aceh and address its concerns directly to the highest levels in Jakarta.

THE CONNECTION BETWEEN PEACE AND DEVELOPMENT

Security is one of the most important preconditions for development. The peace process was critical not only for the post-tsunami reconstruction but also for the longer-term economic development of Aceh. The extended period of conflict destroyed significant infrastructure and curtailed new projects and construction. Unstable environments deter investments and make normal economic activity impossible. Conflict and social unrest also slowly undermine the social infrastructure and lead to a decay in morals, encourage criminality and can lead to all forms of extreme behaviours.

With that said, recovery from both the long-term conflict and the impacts of the tsunami will take years of sustained efforts. Critical to this is the re-integration efforts for both victims of the conflict as well as combatants. Their opportunities to enter into new and productive capacities will be a major step forward in securing Aceh's long-term peace. Furthermore, the reconstruction of houses and infrastructure ruined by the tsunami needs to be embedded within longer-term programmes benefiting all affected citizens of Aceh. As the situation is now, there is a risk of creating tensions and jealousies between people benefiting from different support programmes. This also applies to people in Aceh who were not direct victims of either the conflict or the tsunami, but of endemic and crippling poverty. Wider-reaching development projects need to redress the imbalance in the flow of resources and ensure that no parts of Aceh are allowed to fall too far behind whilst others prosper.

The European Community and European Union member states will continue to support the peace process, including the re-integration of former GAM members. They see the longer-term importance of supporting GAM's transition to a viable, non-violent political party. In the short and mid–term, this involves investing in capacity-building for the new governor and civil administration in Aceh. The gulf between running a long-distance separatist movement and effectively governing a large province is massive. A large part of the future success of the peace process rests upon the availability of administrators and civil servants capable of the practical aspects of governance. This includes training for police in basic policing skills, including community policing and human rights, and support for civil society groups to improve their effectiveness and contribution to the growing democracy in Aceh. Beyond development and capacity-building, most people in Aceh are most concerned with their economic well-being. In the longer term, the best way the international community can support peace in Aceh is through trade, investment and job creation.

THE FUTURE OF GAM

One of the final points to discuss in this chapter is the future of GAM. Obviously, the organization has made massive changes to become part of the political process in Aceh. While it is easy to see the election of Irwandi Yusuf as governor as a positive step that will likely prove beneficial to the peace process, there is still much room for things to backslide. First, it is important that under his leadership there is an increase in trust between the people of Aceh and the powers in Jakarta. The governor and subsequent officials need to establish a lasting trajectory of good governance in accordance with the letter and spirit of the Helsinki MoU. This trust is especially needed as international focus and resources shift away from Aceh, and lasting peace becomes a product of the relationship between Aceh and the rest of Indonesia.

Another major issue is the re-integration of former GAM members into productive roles in mainstream society. While this has clearly happened at the higher political levels and can also be seen in the role that former GAM members played in the BRR, there are many former combatants who need to be assured that they can and need to benefit from the cessation of hostilities. Feelings of economic and social disenfranchisement can play a major role in creating feelings of isolation and disaffection amongst former GAM members. This has been exacerbated by the large amounts of aid money that went to Aceh following the tsunami, and common perceptions about how this money was being distributed. As of 2007, there were an increasing number of cases in which former GAM members were demanding more of a share in reconstruction contracts and employment, and indications that, if not addressed, things could descend into lawlessness in some areas. Because of this, it is essential not only that senior GAM figures become involved in the political process, but also that efforts are made to ensure that the rank and file are not excluded from the development of Aceh. In post-conflict situations, there is always the potential for movements to split into factions, which can undermine the central lines of authority in Aceh and lead to regional tensions. Clearly, this cannot be allowed to happen, and efforts need to be made so that there are equitable plans for longer-term development and re-integration.

CONCLUSION

In this brief chapter, I have drawn from my personal experiences heading the AMM to discuss some of the reasons why the peace process in Aceh was relatively successful in its first three years. The real credit for this achievement rest with various stakeholders in Indonesia, for without their efforts, vision, and pragmatism nothing would have been possible. It is notable that the resolution of the situation in Aceh was greatly facilitated by international partners who provided support during the rounds of negotiations and monitored progress on the ground. There are a number of lessons to be taken from this experience. First, the various local parties must not only have enough will to come to the table and make difficult concessions for a greater good, they must also have the will, discipline and political abilities to rein in potential dissenting voices amongst their respective constituents. Second, a powerful mix of regional powers (ASEAN) and other international partners (the EU) can bring a wide range of resources and influences to bear that has both international legitimacy and local and regional sensitivity. Third, once the agreements have been signed, it is imperative to have rapid and well coordinated actions that follow through on commitments, and an objective party to oversee the process. The IMP and, later, the AMM were able to serve effectively in these capacities because of the strength and flexibility of their mandate.

Three years after the signing of the peace agreement, fighting has stopped, weapons have been decommissioned, and Aceh has begun the long process of establishing its own local political framework and of engaging with the Indonesian Government. This is best exemplified by the people's response to the first major elections in Aceh in 2006, with their election of a former GAM member to the high post of governor. The acceptance of this result by other forces within Indonesia was a major benchmark in Aceh. However, there is still reason to be wary, as peace is fragile and tension and mistrust can often run deep in such situations. It is important that the people in Aceh continue to be supported by the international community in building a stable and prosperous province. It is also important that the changes made by the Indonesian Government are recognized and supported internationally. It is my hope that the experiences in Aceh will provide a model for managing regional autonomy and at the same time preserve wider national stability, making all in Indonesia better off for it. I also hope that the experiences in Aceh are noted by all involved in conflict resolution and that some of the operational features that worked become part of how conflicts are approached in the future.

Notes

1. My comments in this chapter are the result of my personal experiences heading the AMM from 2005 until 2006 and should be taken as such. This is not meant to be an academically grounded discussion, rather a reflection of some of the key points that I believe contributed to the successes of the peace process.
2. Information taken from the European Union website glossary of EU missions: <http://europa.eu/scadplus/glossary/european_security_defence_policy_en.htm>.

11

JUSTICE AND THE ACEH PEACE PROCESS

Leena Avonius

Justice is a tricky word. Everybody makes claims for it and in its name, but very few would be able to explain exactly what it means. This is not because of ignorance, but rather due to the wide scope the term is assumed to cover, and the ambiguities attached to it in its everyday use. Justice — or *keadilan* in Indonesian — is one of the most common terms used when discussing post-conflict processes in Aceh. The absence of justice or the failings of the justice system are seen as major problems in post-conflict reconstruction — and these are addressed through programmes aiming to improve people's access to justice in Aceh. Justice is the key word for local civil society groups that seek to improve its realization and, now and then, to bring their own forms of justice to people in the villages. And justice is what is at stake when the victims of conflict lament that, despite all of the good promises, none of the terrible wrongs they experienced during the conflict have been made right. They are still waiting for justice.

In this chapter, I discuss the question of justice in Aceh's post-conflict reconstruction and peace-building. I argue that to maintain sustainable peace in the territory, it is important to have the widest possible consensus on what is understood by "justice" in post-conflict Aceh. So far, much of the justice talk in the Aceh peace process has been limited to transitional justice issues. Post-conflict reconciliation takes place in relation to institutionalized forms of justice such as human rights courts and truth and reconciliation commissions that seek to provide justice and reconciliation between victims and perpetrators. However, the process of peace-building and reconstruction must also be sensitive to Acehnese views on how a just society should be constructed. This includes talk about social justice, the rights and obligations of members of a society towards each other, as well as the relations between the state and its citizens.

The following will focus on institutionalized forms of justice. I first give a brief overview of the situation in Aceh, then examine how the two signatories of the Helsinki Memorandum of Understanding (MoU) — the agreement that ended three decades of conflict in Aceh in August 2005 — defined the needs and mechanisms of justice in post-conflict Aceh. After discussing briefly the prevailing legal pluralism in Aceh, I examine how the transitional justice mechanisms — amnesties, a human rights court and a truth and reconciliation commission — have been prepared and used in post-conflict Aceh.

LEGACY OF CONFLICT

Aceh has been a scene of violence for a long time. In fact, since the 1870s, Aceh has seen much less peace than armed struggle, fighting and human rights violations. The latest conflict, which began in 1976 and ended in August 2005 with the signing of the peace agreement between the Indonesian Government and the Free Aceh Movement (GAM), took the lives of at least 15,000 Acehnese. Hundreds of thousands of others were either forced to flee their homes or endure similar ordeals. Indeed, as some have stated, practically the entire population of Aceh — over four million people — were directly or indirectly affected by the conflict.

There is still no exhaustive data available on all the damage, destruction and injustices the conflict caused in Aceh. During the conflict years, it was impossible to collect such data, though some non-governmental organizations did their best to document incidents reported to them by victims and their families. Some of these data were lost in 2004 when the Indian Ocean tsunami destroyed NGO offices in Banda Aceh. However, for example, the Acehnese NGO Coalition for Human Rights registered over ten thousand cases of serious human rights abuse before the Helsinki MoU.[1] Rather than accurate information on the number of cases, this type of NGO data offers a good general picture of the patterns of past violence in Aceh: most commonly mentioned are extrajudicial executions, torture, arbitrary arrest, disappearances and sexual violence.

If human rights NGOs catalogued the forms of injustices that took place during the conflict in Aceh, other agencies tended to focus more on the consequences the atrocities have had on the lives of people. For the purposes of post-conflict assistance programmes, there have been a number of data-collecting efforts in recent years. The Aceh Reintegration Agency (Badan Reintegrasi Aceh, or BRA) had, by the end of 2007, registered tens of thousands of people who were eligible for one or more forms of assistance that the BRA provides to both those who had participated in the conflict and the civilian victims of conflict. The BRA's criteria are largely based on verifiable consequences of violence, such as the death or disappearance of a family member due to the conflict, a handicap caused by the conflict, and house or other property destroyed in the conflict.[2] The BRA criteria has been criticized, because some forms of violence — particularly sexual violence, but also other forms of physical and mental violence — do not necessarily leave evidence that can easily be verified years after the incident.[3]

Some idea on how widely and severely the Acehnese population was affected by violence can be attained from the survey on the psychosocial needs of conflict-affected communities, conducted by a Harvard University medical team and the International Organization for Migration.[4] The survey concludes that communities in Aceh are highly traumatized due to conflict-related violence. Respondents suffered particularly from depression (33 per cent), anxiety (48 per cent) and post-traumatic stress disorder (19 per cent).[5] The survey not only confirms NGO information on the patterns of violence during the conflict years, it also includes further categories such as extortion, forced labour and public humiliation.[6]

The sad legacy of conflict can be seen in some communities in Aceh today. The life histories of victims carry the signs of how much injustice they have encountered, as well as the hardships they continue to experience even today. Many women have been forced to raise their children as single mothers because their husbands were either killed or simply disappeared in the conflict. Furthermore, in some cases, men who have survived the conflict are so severely traumatized that they can no longer make a meaningful economic contribution to the household. The violence experienced during the conflict leads to social problems — as is apparent from this story told by a forty-year-old Acehnese woman:

> Early one morning in 2002 the military came to our house. They threatened me with a gun and told me to wake up my husband. Together with several other men in the village he was taken away by the military. I looked for my husband for days but in vain; I have not received any information about him ever since. I have three school-age children whom I now have to bring up on my own. My own health is poor due to all these shocking experiences, and some years ago I was partly paralysed. I was forced to sell our house in order to buy medicine for myself. I cannot work much due to my poor

health and I cannot afford to pay for my children's school fees. Now I see houses being rebuilt for conflict victims, but even though I also lost my home due to the conflict, I am not eligible to get a house.[7]

THE HELSINKI MOU AND JUSTICE ISSUES

The two peace negotiation teams in Helsinki clearly acknowledged the past atrocities in Aceh and how human suffering and poverty had become a part of everyday life for the Acehnese. This is evident in several articles in the MoU that address human rights and justice issues, and create mechanisms for handling them. These include, for example, Article 2.2, which states that "a human rights court will be established for Aceh" and Article 2.3, which determines that the Indonesian Commission for Truth and Reconciliation will establish a Truth and Reconciliation Commission (TRC) for Aceh, and that this TRC will have the task of formulating and determining reconciliation measures. The third part of the MoU concerns amnesty; it outlines government reintegration assistance for those who have participated in GAM activities and for amnestied political prisoners, as well as assistance for conflict-affected civilians. While these articles deal mostly with the past, the preceding part of the MoU outlines how sociopolitical justice would best be ensured in Aceh in the future through guaranteeing full political participation, economic development, the rule of law and adherence to international covenants of human rights.

Even though the Helsinki MoU includes many articles that deal with justice issues, there appears to have been very little discussion about justice during the peace talks in Helsinki. Edward Aspinall points out in his assessment that few participants have any recollection of how justice issues were dealt with.[8] On the other hand, there was no resistance from the two parties when the mediating organization, Crisis Management Initiative (CMI), introduced transitional justice mechanisms to the draft document. Justice issues may have been overshadowed by security issues, as evident from the number of published accounts of the Helsinki peace talks.[9] However, issues related to social and political justice were not missing.

The Helsinki MoU is an informal document, but most of its articles were given a legal basis in the Law on the Governance of Aceh (LoGA), which was passed in July 2006 by the Indonesian national parliament, the DPR. The implementation of the MoU, through the LoGA and otherwise, has similarly been characterized by a lack of extensive debate on justice issues. Mostly, justice has been brought up in relation to transitional justice mechanisms, particularly the plans to establish a truth and reconciliation commission.

As will be discussed below in more detail, these debates have tended to stress the legalistic and technical aspects of transitional justice processes. This has at least partly led to a situation where justice in post-conflict Aceh is understood in a narrow sense as "legal justice", while "social justice" has received less attention.

JUSTICE AND INJUSTICE IN ACEH

Before examining how justice issues have been discussed in post-conflict Aceh, it is necessary to explore more generally the conception of justice in Aceh. As in other Indonesian regions, in Aceh justice is understood within a context of normative pluralism. What is just in any particular situation should be examined through three normative frameworks: national laws, Islamic law and *adat*. This does not mean that all these frameworks are actively used in all situations. It is possible that in some situations one of them may be considered to be more relevant or appropriate than the others.

It is a generally shared assumption in Aceh that *adat* and Islamic law are congruent, and that one cannot really exist without the other. Though inseparable, they are not the same. Sometimes, *adat* and Islamic law appear to be acceptable alternatives and are considered to provide equally just results — even though if followed alone they would lead to very different results. The most common example is found in the inheritance practices that combine *adat* and *syariah*. I have not studied this issue in detail, but my discussions with Acehnese women have given me the impression that some of them have inherited land that was the property of their mothers, suggesting that *adat* practices have been followed. In other families, Islamic inheritance practices have been followed, leaving daughters to inherit half of what their brothers get. Such conflicting norms can lead to family disputes, but the incongruence between *adat* and religion is often solved by making use of the Islamic practice of *hibah*, a gift from parents to children when they marry. Daughters receive their *adat* land as *hibah*, while formal inheritance follows Islamic *syariah*.[10] *Adat* practices vary in Acehnese regions: some families are more inclined to follow *syariah*, whereas other families prefer *adat* when decisions need to be made.

John Bowen[11] has questioned the suitability of justice theories created by such Western thinkers as John Rawls in pluralist societies like Indonesia. He argues that, while Rawls assumes the existence of a "political conception of justice", in places like Aceh or Indonesia, justice is "public and also Islamic". Bowen does not discuss Rawls's political conception of justice extensively, but merely states that it is based on Rawls's own experiences in American society and "conjectural" theorization. He concludes that justice in Aceh always remains to be understood and expressed through Islamic reasoning, and that it is impossible to merely extract "public reason" that would be divested of its religious elements. Bowen's criticism of Rawls appears to be based on cultural relativism: he states that his empirical evidence from Aceh shows that universalistic conceptions of justice are not applicable, for justice issues are always argued through religious concepts and reasoning.

While I agree with Bowen's argument that justice theories like Rawls' tend to be highly idealistic and can be easily challenged through empirical evidence of the complexity of everyday life, I think that he is overemphasizing the role of Islam as the primary source of justice in Aceh. In the post-conflict debates on human rights abuses in Aceh, it is evident that, rather than Islamic law, it is secular conceptions of justice based on Indonesian national laws and international human rights principles that are taken to be the foundations of public reasoning in these cases. Sometimes, *adat* is offered as an alternative framework for national laws, but explicit references to Islamic law are rarely made, even when concepts emerging from it are used.[12] Islamic concepts of justice are not used exclusively and, in relation to human rights violations, universalistic concepts may take priority.

The chosen normative framework is highly relevant in cases of human rights abuses, as the understanding of what kind of punishment is sufficient to provide justice differs greatly in Acehnese *adat* and Indonesian national laws. For example, according to the Indonesian Criminal Code, in cases of serious maltreatment committed with premeditation, the appropriate punishment is a maximum of twelve years imprisonment. Acehnese *adat* tends to stress a reconciliatory approach; in most of the cases that I have followed, the typical resolution has required the perpetrator to cover all of the victim's medical expenses, pay compensation either in cash or in kind, and arrange a communal *peusijuek* ritual to mark the settlement of the case.

Among the few instances of a practice based on Islamic law that is being used in addressing post-conflict grievances is *diyat*, compensation given by the provincial authorities to the family of persons killed in the conflict. This practice has incurred extensive criticism in Aceh. *Diyat* practice was created in 2002 by the then vice-governor Azwar Abubakar. According to *diyat*, the family of the victim receives an annual subsidy from the government; in post-conflict Aceh, the sum has been three million rupiah per year. Civil society activists advocating justice for past human rights abuses, as well as family members of victims, have particularly criticized and turned down *diyat*, claiming that its purpose is to maintain impunity. Their criticism does seem to be grounded, since some government officials and Islamic scholars have suggested that the acceptance of *diyat* is a sign that the perpetrator has been forgiven.[13] Another problem with *diyat,* like any other form of cash assistance in Aceh, is that it has often been paid to people who have not been conflict victims.

Post-conflict justice issues in Aceh are further complicated by a general lack of knowledge on justice and rights that enables the abuse of power by justice experts and government officials. The awareness of how justice systems function or even of basic rights is very low amongst ordinary villagers in Aceh, whether the question is about national law, *syariah* or *adat*.[14] Marginalized groups, such as conflict victims, are especially poorly informed about justice issues. For example, more often than not, people in villages are unaware that they can bypass the village leaders such as *keucik* and *tuha peut* and take their complaints directly to the district court; that is, that they themselves have the right to decide whether to use the national justice system or *adat* justice. While the public debates on human rights abuses that I referred to above indicate that the public awareness of justice issues is improving in some areas, it will take years of campaigning and public education to empower villagers. Furthermore, it is not only ignorance that prevents villagers from seeking justice but also lack of trust. This is largely because national courts are generally seen as corrupt. During the conflict years, the national courts were not trusted by Acehnese with GAM connections, as such courts were seen to be politically biased towards the interests of the central state.

To summarize what has been discussed in this section, I would say that the domain of justice in post-conflict Aceh is complex and filled with ambiguities. There is a commonly held view that legal pluralism prevails in Aceh and that the normative frameworks in use are Indonesian national legislation, Islamic law and *adat*. At a general level, these normative frameworks are seen to be supportive of each other and to a large extent congruous; in practice, the normative frameworks offer different options on how to provide justice in any particular case. These differences need to be negotiated and can become sources of difficult disputes. The low level of public awareness of justice systems and the lack of confidence in them make the sphere of justice in Aceh ambiguous and its results in many cases unpredictable. There can be many discrepancies between how ordinary people in Aceh see justice and how justice experts of any justice system or power-holders in communities understand and practice it.[15]

If there is ambiguity regarding the sources of justice and their use in post-conflict Aceh, nobody in Aceh would deny that severe injustices took place during the conflict years or that they exist post conflict. The UNDP report on access to justice in post-tsunami and post-conflict Aceh lists the most commonly articulated justice grievances in post-conflict areas. These include past human rights violations, theft and destruction of property, destruction of livelihoods, government and village-level corruption, displacement due to the conflict, land disputes, violence against women and inheritance disputes.[16] Many of these grievances require formal justice procedures for their settlement, while other problems such as corrupted government practices and lost or reduced livelihood opportunities cannot be addressed in the courtroom. They require improved social practices, that is, reforms congruent with Acehnese ideas of a just society.

TRANSITIONAL JUSTICE AND DEALING WITH THE PAST

As discussed above, the Helsinki MoU proposed three transitional justice mechanisms to be used to deal with the past in Aceh: amnesties, a human rights court (HRC), and a truth and reconciliation commission (TRC). Three years after the signing of the MoU, the amnesties have more or less been completed, while only very initial steps have been taken towards the implementation of the HRC and TRC articles. In this section, I discuss the problems hindering the establishment of these two bodies as well as possible solutions to the problems encountered.

Before discussing HRC and TRC, a few words should be said about amnesty measures in Aceh. As the MoU had stipulated, the Indonesian Government granted amnesty to "all persons who have participated in GAM activities" through Presidential Decree no. 22 of 2005. On the basis of the decree, over 2,000 GAM prisoners were released, most of them at the end of August 2005.[17] The implementation of the amnesty agreement has proceeded without major problems. The Aceh Monitoring Mission (AMM) was authorized to rule on disputed cases; it took part in resolving some sixty disputes between GAM and the Indonesian Government in 2006. Of these cases, the release of all but eleven prisoners was agreed to by the Indonesian authorities. Even though Malik Mahmud, the GAM signatory to the Helsinki MoU, agreed to this settlement, negotiations are still continuing over some of the eleven cases, including persons who were sentenced for their involvement in the Jakarta Stock Exchange bombing in 2000 and other serious crimes.[18] However, their amnesty has not been refused on the basis of serious crimes (the Presidential Decree on Amnesty does not stipulate that amnesties could be refused to those who have committed human rights abuses) but because their crimes were not considered to be related to GAM activities. Recent negotiations have concerned remissions and a possibility that these prisoners could be transferred to serve their sentences in Aceh prisons.[19]

Although amnesties have proceeded well, this cannot be said for the implementation of the HRC article of the MoU. The only substantive step that has been taken towards the establishment of a human rights court for Aceh, since it was agreed in the MoU, has been the repetition of this promise in the Law on the Governance of Aceh (LoGA, Law 11 of 2006). The LoGA stipulated that an HRC should be established by July 2007 (Art. 260), but this timeline was unrealistic, considering the political sensitivities of the post-conflict situation in Aceh and the general weakness of the judicial system in Indonesia.

More serious consequences to the establishment of an HRC in Aceh are brought about by Article 228 of the LoGA, which states that the HRC's mandate will be limited to rule on cases taking place after the passage of LoGA (July 2006). Such a stipulation clearly indicates that the HRC in question is not meant to be a transitional justice mechanism, for it would not handle the past atrocities at all.[20] Instead, its purpose would be to ensure that, in the future, human rights violations would be taken to court for prosecution. Defined in this way, the HRC's mandate overlaps with the mandate of the permanent human rights court in Medan, which, according to the Indonesian law on human rights courts (Law 26 of 2000), has the authority to process gross human rights violations that have taken place in North Sumatra, Aceh, Riau, Jambi and West Sumatra after November 2000. It is not surprising that the Department of Law and Human Rights has recently recommended that the HRC for Aceh as meant in the MoU and LoGA should be established as a part of Medan's permanent human rights court, and would be placed at the general court in Banda Aceh.[21] If the HRC for Aceh is established according to this recommendation, it will be of the utmost importance to clarify whether the court in Banda Aceh will deal with incidents that have taken place after November 2000 (in accordance with the law on human rights courts) or only after July 2006 (as according to LoGA). This is important, since it is exactly during this time period that most of the severe human rights abuses took place in Aceh.

Even though the intentions to establish a retributive justice mechanism to look into past atrocities were watered down by the LoGA, Indonesian human rights legislation does offer another mechanism to deal with them. The law on human rights courts states that an ad hoc human rights court can be established to deal with gross human rights violations that have taken place before November 2000 (Art. 43/1). An ad hoc court can be established if there is a recommendation from the DPR, with approval from the president of Indonesia. So far, there has been no initiative in Aceh to lobby the national parliament for an ad hoc court. Previous efforts to deal with past atrocities through ad hoc human rights courts in Indonesia have been discouraging: there have been two ad hoc courts for the massacres in Tanjung Priok in 1984 and in East Timor in 1999, and both courts upheld impunity rather than provided justice for the victims. A further obstacle to establishing an ad hoc court for Aceh was that, amongst the members of the national parliament in 2007–09, there were several who vocally criticized and resisted the peace talks in Helsinki. Any suggestion to establish a human rights court would have met strong resistance in the parliament. This situation has not changed much since parliamentary elections in 2009.

My interviews with conflict victims in Aceh largely correspond to the UNDP's assessment on access to justice.[22] Victims expressed a strong desire for justice; they believed that perpetrators of past atrocities deserved punishment. However, they were also painfully aware of the improbability of justice ever taking place. On the one hand, victims have little trust in courts that are viewed as corrupt and dysfunctional; on the other hand, victims acknowledge that it would often be impossible for them to even identify the perpetrators.[23] Even if established, human rights courts would handle only the most serious cases. HRCs in Indonesia seek punishment for injustices and wrongdoings that fall under the universally recognized category of gross human rights violations of genocide and crimes against humanity.[24]

Since 2005, there have been efforts to push ahead with the establishment of a truth and reconciliation commission in Aceh, but unlike in the HRC process, there are legal hindrances in its way. Both the MoU and the LoGA determined that the TRC in Aceh should be established within the national TRC system, as meant in Law 27 of 2004. In December 2006, however, the Indonesian Constitutional Court abolished that legal framework by annulling the law in question. This was somewhat shocking to the Acehnese civil society activists who had started to promote the TRC in Aceh. Some of them even speculated that the Constitutional Court's motive for making such a decision was to prevent truth-seeking efforts in Aceh.[25] While there is strong resistance to the examination of past atrocities and, particularly, the role of the Indonesian military in them, revoking the TRC law by the Constitutional Court was not unequivocally negative. Law 27 of 2004 had been taken to the Constitutional Court by Indonesian human rights advocates who were concerned that some of its articles were extremely problematic in light of human rights principles.[26] The court decision sided with them, but instead of revoking only the problematic articles, it did away with the whole law.

The legislative vacuum left by the Constitutional Court decision is indeed problematic for the implementation of the TRC articles of the MoU and the LoGA. Some suggested that Aceh should establish its own truth and reconciliation commission through a regional regulation (Qanun), but such a commission would have a very limited mandate and could not guarantee that perpetrators residing outside Aceh could be summoned to testify. In 2007, several seminars were jointly organized by the Aceh provincial government, the national human rights commission, local civil society organizations and some international agencies active in transitional justice issues to discuss options and to share views and opinions on the TRC. These seminars as well as of meetings between local civil society representatives and Aceh's provincial office responsible for legal issues, concluded that the most viable alternative for Aceh would be to wait until a new national law on truth and reconciliation commissions is instated. A new draft bill has been prepared by the Department of Law and Human Rights, and it has been included in the national legislation programme (*Prolegnas*) of 2008. It is, however, unlikely that the national parliament would be able to pass the bill this year, due to its controversial nature and the upcoming elections next year.

All in all, it looks more likely that in Aceh the TRC will be given priority over the HRC. It has become very clear in public statements made by Aceh's leadership, the provincial parliament members and Acehnese civil society representatives that there is political will in Aceh for establishing a TRC.[27] There are some people in Aceh, usually with links to the alleged perpetrators of past human rights abuses, who are of the opinion that Aceh needs neither a TRC nor an HRC. They tend to see any queries into past atrocities as a threat to the security situation. Nonetheless, the vast majority of people I have interviewed in Aceh support the establishment of transitional justice mechanisms, even though their ideas on the forms and working methods of these mechanisms differ.[28]

The Helsinki MoU and the LoGA provide very few guidelines on how the TRC in Aceh should look. Some have criticized the MoU for not providing more detailed provisions on the HRC and the TRC.[29] I would say, however, that, considering that the peace talks in Helsinki did not involve any expertise on transitional justice issues, the general phrasing of these mechanisms in the peace agreement is actually more of a blessing than a problem. After all, the only detail that was included in the MoU concerning the TRC — the linking of the Aceh TRC to the national truth and reconciliation mechanism — has become a major obstacle for the establishment of a TRC in Aceh. Lessons learned from other peace processes show that general outlines tend to be better than detailed provisions, since in the latter case political bargaining may handicap the entire system. Also, the persons at the negotiating table may not possess a good understanding of the complexity of the needs of various victim groups, particularly the special needs of women, children and minorities.[30]

Despite all the hindrances, the establishment of a TRC in Aceh in the next few years is still possible. There is strong support for it there. And at the national level, the recent release of the report of the Truth and Friendship Commission between Indonesia and East Timor suggests that attitudes towards transitional justice mechanisms might be changing amongst political leaders. The report admitted, for the first time ever, the involvement of Indonesian security forces in human rights violations in East Timor in 1999. Though the commission failed to take the process a step further by recommending prosecutions, the report is a positive sign for Aceh and other post-conflict regions in Indonesia.

Acehnese civil society has been proactive in drafting a model for Aceh's TRC. They have formed a coalition, the KPK (*Koalisi Pengungkapan Kebenaran*, Coalition for Truth Recovery), which has prepared a background paper with information on past violations that a TRC should investigate; previous truth-seeking efforts in Aceh; legislative, structural and cultural frameworks for the establishment of a commission; and an initial model for a TRC in Aceh.[31] The draft indicates that, according to the KPK, the commission's work should be based on Islamic ethical values stressing the importance of truth, justice and reconciliation,[32] and that the commission should make use of Acehnese *adat* practices such as *diyat, sayam, suloeh* and *peumat jaroe* in its reconciliation efforts.[33] Thus, the TRC in its work should manage to combine the three normative frameworks that are acknowledged in Aceh. It is still too early to say whether and how this would work in practice. While optimists say that using local *adat* would guarantee community support for the TRC, pessimists fear that stressing local traditions extensively may lead to a situation where the commission overemphasizes "reconciliation" while ignoring the "truth".[34]

A further issue causing differences in opinion is how far back in history the TRC should reach in its truth-seeking efforts. Those who support the narrowest time frame suggest that the commission should limit its examinations to the period after the first military operation was launched in 1989 by the New Order regime. Others would like to include the 1970s in the examination, arguing that the early years of the Free Aceh Movement cannot be ignored. Yet others insist that the patterns of violence in Aceh cannot be fully understood if the Islamic rebellion of the Darul Islam movement in the 1950s is left out. All of these opinions have valid points and, whatever the chosen scope will eventually be, it is important that this, as well as other elements of the commission, is decided on the basis of a wide public debate on TRC in Aceh. So far, the discussion on the TRC in Aceh has been limited to the provincial elite, whether in government or civil society, leaving the victims on the margins. As truth commissions in general should be based on the needs of the victims, the "elitism" of the process is a concern that needs to be addressed.[35]

CONCLUSION

Discussions on justice issues in post-conflict Aceh tend to be limited to transitional justice mechanisms, which run the risk of creating unreasonably high expectations for the HRC and TRC. While the conflict has caused an enormous amount of suffering in Aceh, causing either directly or indirectly the misery and poverty people are forced to live in, these institutions can do very little for building a just society in the future. Aceh's situation differs little from many other post-conflict societies, where a balance needs to be struck between dealing with the past and planning for the future. Transitional processes have also been ongoing throughout Indonesian regions since the late 1990s, and are usually referred to by such terms as *reformasi*, or regional autonomy and decentralization. After some years of delay, Aceh has now joined these reform processes and, like other Indonesian regions, it will need to develop its own recipes for *reformasi* and social justice. Indonesia, with its hundreds of sociocultural systems and traditions, certainly provides evidence that one size and model does not fit all.

The Helsinki peace agreement indicated that three transitional justice mechanisms should be used in dealing with the past in post-conflict Aceh, but it largely left the design of these mechanisms open. While the implementation of the amnesty article of the MoU has been generally unproblematic, the processes to establish a TRC and an HRC have encountered major obstacles, both in terms of a lack of political will and inadequate or unclear legal frameworks. Assessing the initial steps of these processes, one gets the impression that the two mechanisms will eventually be established in Aceh, though it may still take several years. Indonesia's previous experiences with ad hoc human rights courts suggest that the outcome of the mechanisms may, however, turn out to be disappointing. The country's human rights track record has improved remarkably during recent years, but impunity for the perpetrators of human rights abuses is still indicated as a major problem both by Indonesian and international observers of Indonesia's justice sector. As I mentioned above, some in post-conflict Aceh have raised concerns that prevailing legal pluralism may offer opportunities for upholding impunity. This does not mean that they would question legal pluralism as such. Rather, it pinpoints the importance of having a general consensus on which the justice system will be used in dealing with any particular type of case, and on how the various systems in use will be combined in the processes of transitional justice. Such a consensus can be reached only through an open, public discussion on justice issues.

Notes

1. Unpublished data from Koalisi NGO HAM. This data include only those cases that were reported to their office. The organization was established in 1998, and thus its information is focused more on the post-Suharto era.
2. The ten-point criteria that were originally agreed by the representatives of the Indonesian Government, GAM and Acehnese civil society organizations on 15 March 2006: 1) a person deceased due to the conflict (*diyat*); 2) a person who became a widow/ widower/orphan due to the conflict; 3) a person gone missing due to the conflict; 4) a person whose house was burnt, broken or destroyed due to the conflict; 5) a person whose wealth was destroyed/broken/lost due to the conflict; 6) a person who was displaced from the place of origin due to conflict; 7) a person disabled or dismembered due to the conflict; 8) a person mentally ill due to the conflict; 9) a person who is physically ill due to the conflict; and 10) a person who lost occupation due to the conflict. The data collection was delayed due to internal problems in BRA, but since 2007 BRA district offices have collected data on conflict victims in villages.
3. Interviews by the author in Aceh in 2007–08.
4. International Organization for Migration (IOM), *Psychosocial Needs Assessment of Communities Affected by the Conflict in the Districts of Pidie, Bireuen and Aceh Utara*, 2006; and International Organization for Migration (IOM), *A Psychosocial Needs Assessment of Communities in 14 Conflict-Affected Districts in Aceh*, 2007.
5. IOM, *Psychosocial Needs Assessment of Communities in 14 Conflict-Affected Districts in Aceh* (2007), pp. 52–53.
6. Ibid., pp. 24–25.
7. Interview by the author in Aceh in 2008. In 2007, BRA introduced an assistance programme for children's education, though it only includes children who were orphaned due to the conflict. Due to budgetary delays, BRA has been unable to provide the promised assistance. Many families interviewed also pointed out that in fact it was unjust for schools to ask for fees, as primary education should be free in Indonesia.
8. Edward Aspinall, *Peace without Justice? The Helsinki Peace Process in Aceh* (Geneva: Centre for Humanitarian Dialogue, 2008), pp. 16–18.
9. Hamid Awaluddin, *Damai di Aceh: Catatan Perdamaian RI-GAM di Helsinki* (Yogyakarta: CSIS, 2008); Farid Husein, *To See the Unseen: Scenes Behind the Aceh Peace Treaty* (Jakarta: Health & Hospital Indonesia, 2007); Katri Merikallio, *Miten Rauha Tehdään: Ahtisaari ja Aceh* (Juva: WSOY, 2006); Damien Kingsbury,

Peace in Aceh: A Personal Account of the Helsinki Peace Process (Jakarta: Equinox Publishing, 2006).

10. Kathryn Robinson, "Gender, Islam and Culture in Indonesia", in *Love, Sex and Power: Women in Southeast Asia*, edited by Susan Blackburn (Clayton: Monash Asia Institute, 2001), referring to Chandra Jayawardena, "Women and Kinship in Acheh Besar, Northern Sumatra", *Ethnology* 16 (1977): 21–38.

11. John Bowen, *Islam, Law and Equality in Indonesia: An Anthropology of Public Reasoning* (Cambridge: Cambridge University Press, 2003), pp. 11–12, 264–65.

12. For example, in the public debate on a shooting incident in Paya Bakong, North Aceh in 2006, in which the Indonesian military shot dead a former GAM member and injured several other persons, *adat* procedures suggested for reconciliation were strongly rejected by the victims and their families, who insisted that in cases of human rights violations only national courts and legislation can be used. For a detailed discussion on the case, see Leena Avonius, "Reconciliation and Human Rights in Post-Conflict Aceh", in *Reconciliation from Below: Grassroots Initiatives in Indonesia and East Timor*, edited by Birgit Bräuchler (London: Routledge, 2009).

13. Aspinall, *Peace without Justice?* pp. 25–26.

14. United Nations Development Programme (UNDP), *Access to Justice in Aceh: Making the Transition to Sustainable Peace and Development in Aceh*, UNDP Report 2007 <http://www.undp.or.id/pubs/docs/Access%20to%20Justice.pdf> (accessed 19 May 2009).

15. The UNDP report (2007) on access to justice in Aceh discusses many such discrepancies in relation to how Acehnese assess justice systems in general and in relation to their own experiences of the functioning of those systems.

16. UNDP, *Access to Justice in Aceh*, p. 35.

17. The extrajudicial arrests and custody explain why the total number of amnesties varies according to source. An Acehnese civil society network, Forum for Justice to Acehnese Political Prisoners (FKTNA), stated that, by September 2007, a total of 1,488 persons were granted amnesty, while the number of amnestied prisoners that had been given reintegration assistance by the BRA was 2,035. The BRA has considered some cases and also provided assistance to some prisoners who have been unable to show any written evidence on their imprisonment (Recommendation 05/SK/FKTNA/IX/2007 by FKTNA; BRA interview in December 2007).

18. Aspinall, *Peace without Justice?* p. 20.

19. Commission on Sustaining Peace in Aceh (CoSPA), press release of CoSPA meeting on 17 June 2008 <http://www.bra-aceh.org/details_cospa.php?bahasa =indonesia&id=501> (accessed 19 May 2009).

20. A good definition of transitional justice as commonly understood is given by Bickford: "Transitional justice refers to a field of activity and inquiry focused on how societies address legacies of past human rights abuses, mass atrocity, or

other forms of severe social trauma, including genocide or civil war, in order to build a more democratic, just, or peaceful future." Louis Bickford, "Transitional Justice," in *Macmillan Encyclopedia of Genocide and Crimes against Humanity*, vol. 3 (New York: Macmillan, 2004).

21. Commission on Sustaining Peace in Aceh (CoSPA), press release of CoSPA meeting on 16 April 2008, Badan Reintegrasi-Damai Aceh, <http://www.bra-aceh. org/details_cospa.php? bahasa=indonesia&id=115> (accessed 19 May 2009).

22. See UNDP, *Access to Justice in Aceh.*

23. Interviews by the author in Aceh 2006–08.

24. The definitions are adapted from and consistent with international standards, as pointed out recently by Suzannah Linton in "Accounting for Atrocities in Indonesia", *Singapore Year Book of International Law* 10 (2006): 1–33. Indonesian law defines "genocide" along the lines of the definition formulated by the Genocide Convention, while the definition of "crimes against humanity" follows the way these have been outlined in the Statute of International Criminal Court (ICC).

25. Interviews by the author in Aceh 2006–08.

26. Two issues were considered to be particularly problematic: first, that some of its articles conditioned the victim's right to receive compensation for amnesty granted to the perpetrator, and second, that the law presented human rights courts and a truth and reconciliation commission as alternatives to each other, even though they are generally seen as complementary mechanisms (interview with a Komnas HAM representative in 2007).

27. In January 2007, Acehnese and Indonesian media reported that elected vice-governor of Aceh, Muhammad Nazar, assured that a TRC would be established in Aceh in 2007, while Mukhlis Mukhtar, the chair of the Law and Governance Section of Aceh's provincial parliament, stated that the decision of the Constitutional Court did not prevent Aceh from establishing a TRC; see Tempointeraktif, "Aceh Tetap Bentuk Komisi Kebenaran dan Rekonsiliasi", 24 January 2007 <http://www.tempointeraktif.com> (accessed 19 May 2009). In November 2007, the governor of Aceh, Irwandi Yusuf, assured the participants of a seminar in Jakarta that his government would support any efforts to establish a TRC; see *The Jakarta Post*, "Sharia Law 'Could Support Aceh Truth and Reconciliation'", 24 November 2007 <http://www.aceh-eye.org/a-eye_news_files/ a-eye_news_english/news_item.asp?NewsID=7715> (accessed 19 May 2009).

28. Interviews by the author in Aceh 2006–08.

29. Aspinall, *Peace without Justice?* p. 18.

30. International Council on Human Rights Policy (ICHRP), *Negotiating Justice? Human Rights and Peace Agreements*, pp. 44, 92, <http://www.ichrp.org/en/ documents> (accessed 19 May 2009).

31. Koalisi Pengungkapan Kebenaran (KPK), *A Proposal for Remedy for Victims of Gross Human Rights Violations in Aceh*, an unpublished working paper (2007). See also Aspinall, *Peace without Justice?*

32. KPK, *A Proposal for Remedy*, pp. 7–8.
33. *Diyat* has already been discussed above; *sayam* is compensation paid from the perpetrator to the victim in cases where "blood has been spilled". *Suloeh* is a peace-building effort in a community, particularly in economic disputes, while *peumat jaroe* refers to handshaking that signifies that a dispute has been resettled (KPK, *A Proposal for Remedy*, pp. 23–25).
34. Interviews by the author in Aceh 2006–08.
35. Priscilla B. Hayner, *Unspeakable Truths: Facing the Challenge of Truth Commissions* (New York: Routledge, 2002).

12

MANAGING PEACE IN ACEH
The Challenge of Post-Conflict Peace Building

Rizal Sukma

INTRODUCTION

Unlike the previous two peace attempts, the Helsinki peace accord reached by the Government of Indonesia (GoI) and the Free Aceh Movement (GAM) in August 2005 appears to have a better chance of bringing an end to the separatist conflict in the Province of Nanggroe Aceh Darussalam (NAD). More than four years after the implementation of the agreement, peace in Aceh was still holding. After the conclusion of peaceful regional elections on 11 December 2006, the overall picture has become even more encouraging. For GAM, its decision in Helsinki to transform itself from an armed insurgency group into a political force within the Republic of Indonesia began to pay off when many former GAM leaders, including the candidate for governor Irwandi Yusuf, won the local elections. For the Government of Indonesia, the fruit of a political settlement to the conflict was evident when Irwandi Yusuf and Muhammad Nazar officially took the oath as the new governor and vice-governor of NAD Province on 8 February 2007, pledging allegiance to the Republic of Indonesia.

The Aceh peace process represents a remarkable example of a peaceful settlement of internal conflict in a democratizing country. However, because the conflict in Aceh is deep-rooted and multifaceted, factors that could derail the peace process have not disappeared entirely. The key challenge now for the Indonesian Government, the new Aceh government and the Acehnese is how to manage the difficult task of post-conflict peace-building in order to ensure that conflict does not recur in the future. This chapter analyses why the Helsinki peace accord, known as the Memorandum of Understanding (MoU), has so far worked well. In addition, the chapter also examines future challenges in post-conflict peace-building efforts. Three issues are critical in this regard: the challenge of governance, the progress of post-tsunami reconstruction and the imperative of a democratic political order in Aceh. Finally, the chapter draws some lessons that might be relevant to the resolution of other conflicts in the region.

THE STATE OF PEACE PROCESS: NOT PERFECT, BUT MOVING FORWARD

Before the Helsinki peace process, there had been two previous attempts at finding a peaceful resolution to the conflict in Aceh. The first attempt, which culminated in the agreement to initiate a Humanitarian Pause in 2000, did not last very long. The second peace attempt, the Cessation of Hostilities Agreement (COHA) in December 2002, lasted for only six months until, in 2003, the Indonesian Government decided to impose a military emergency status in Aceh and launched a massive military operation. When the GoI and GAM concluded the Helsinki peace accord in August 2005, many expressed their worry that this latest peace attempt would also suffer a similar fate as the other two attempts. However, more than thirty-six months after the implementation of the peace agreement between the GoI and GAM, the peace process in Aceh continues on the right track.

Indeed, the implementation of the terms of the peace process as envisaged in the MoU has been relatively smooth (see the account provided by Feith in this volume). Violence has subsided significantly across Aceh since the agreement was signed. From late January until the end of August 2005, for example, incidents of violence increased steadily, causing 179 deaths and 172 injuries.[1] However, since the signing of the MoU, incidents of violence have drastically declined. What is more encouraging is the fact that the implementation of the terms of agreement has been completed according to the timetable agreed upon by both sides. For example, the process of arms decommissioning and the demobilization of GAM and TNI/police troops was completed almost without major obstacles. The government has delivered on its promise to grant amnesty and pardons for political prisoners. The GoI has also delivered its obligation to implement the re-integration programmes in accordance to the MoU, especially in providing economic incentives to former GAM combatants. This progress has certainly improved confidence and trust, and allowed the peace process to move forward.

On the question of Aceh's status and place within the Republic of Indonesia, differences between the central government and Aceh were finally resolved, albeit temporarily, with the passing of the Law on Governing Aceh (LoGA) by the Indonesian Parliament (DPR) on 11 July 2006. The LoGA provides not only the detailed framework for engaging the clauses of the MoU, but also serves as the foundation for rebuilding and governing Aceh as an autonomous province within the Republic of Indonesia. More importantly, the passing of LoGA by DPR clearly indicates a broad consensus among competing political forces in Jakarta — including those who were initially opposed to the peace deal such as the Indonesian Democratic Party–Struggle, or PDI–P, and sections of the armed Forces — on the necessity of compromise for the sake of peace in Aceh. In other words, the LoGA should be seen as a testimony to the presence of a constituency of peace in Jakarta.

On the role of external actors, initial suspicions among some circles in Jakarta that they would not act as neutral parties have not been vindicated. On the contrary, the role of external actors, especially the AMM, and also the support by international agencies in helping the implementation of the peace deal, has reinforced the chance for success. The work of the AMM has been widely praised within Indonesia by the GoI, GAM and society at large. It has also demonstrated itself as a highly professional team that stands by the principle of impartiality. A survey by IFES shows that 97 per cent of Acehnese are satisfied with AMM's performance.[2] The implementation of the peace accord is also greatly helped by support from international agencies. The World Bank and the International Organization for Migration (IOM), for example, have worked in partnership with the government to assist the re-integration programme of former GAM combatants into mainstream society, through various economic facilitation programmes.[3]

The most important indicator of peace in Aceh has been the return of a sense of normalcy in ordinary peoples' lives. First and foremost, this sense of normalcy has been demonstrated by the ease of travel throughout Aceh. Economic activities that functioned primarily in medium and big towns during the conflict period have now also returned to small towns in subdistricts and villages. Farmers can now freely go to their farmland without any fear of being caught in a shootout between the army and GAM rebels. Indeed, a visitor observes:

> What is my dominant image in Aceh? It is of Acehnese celebrating the return
> of life toward normalcy: streams of students walking on the streets near

campus; tents being replaced by more permanent housing, shops, open-air markets and cafes flourishing again. It is of traffic jams returning to the main thoroughfares of Banda Aceh, and of people travelling freely again on roads throughout the region. It is the first harvest of rice in fields destroyed in December 2004, and people remarrying and starting new families.[4]

The peace process in Aceh has been consolidated even further by the completion of peaceful local elections on 11 December 2006. Despite some political manoeuvrings and tensions among the competing forces, as well as intra-party differences within GAM before the elections, the race for political power has not disrupted the peace process. The victory of several GAM-affiliated candidates, both at provincial level and in eight districts/cities, clearly marked the beginning of a new era in Aceh. The central government leaders have also pledged that they will support and work closely with the Aceh administration. As the swearing-in ceremony of the new governor and vice-governor went on without any hindrance, there is ground to be optimistic that the prospects for a better and peaceful Aceh is not beyond reach.

WHY IS PEACE STILL HOLDING? THE NATURE OF HELSINKI ACCORD[5]

Why is this latest Aceh peace process still holding and, indeed, consolidating? The answer to this question lies primarily in the nature of the peace agreement itself. The MoU, which constituted a breakthrough in the long process of resolving the separatist conflict in Aceh, registers a set of qualities that make the current peace process more attainable. It addresses many (if not all) key issues in a relatively comprehensive manner, reflects the willingness of both sides to compromise and avoid a zero-sum position and, above all, provides a creative formula for dealing with the most difficult issues that the previous peace talks had not been able to resolve, namely the issue of GAM's existence, its demand for independence and the question of possible resistance to the peace deal, either from within GAM or the Indonesian military.

The MoU is comprehensive for four main reasons. First, unlike the Humanitarian Pause and COHA, which focused on the establishment of a cessation of violence on the ground, the MoU represents the first attempt to achieve a comprehensive political solution to the conflict. Second, the signatories to the MoU for the first time utilized disarmament, demobilization and re-integration (DDR) as an integrated framework for the peace process. The DDR issue, which constitutes an important component for any resolution of conflict, was not addressed adequately in the previous peace agreements.[6]

Third, the MoU addresses, though not fully, a wide range of issues: legal issues, governance, Aceh's status, economic incentives, political participation, human rights and reconciliation. Fourth, it also provides mechanisms for implementation, including institutional arrangements and a mechanism for dispute settlement.

With regard to disarmament and demobilization, the agreed steps provided the basis for ending the deep-seated hostility between the TNI and GAM, and built confidence and trust between the two parties. Both sides agree that GAM will undertake "the decommissioning of all arms, ammunition and explosives held by the participants in GAM activities"[7] and the GoI "will withdraw all elements of non-organic military and non-organic police forces from Aceh"[8] to be conducted "in parallel with the GAM decommissioning".[9] The MoU stipulates that GAM will demobilize all of its 3,000 military troops. It was also agreed that GAM will hand over 840 arms and the TNI will reduce the number of its troops stationed in Aceh to 14,700. The MoU provided a clear timetable for both sides to complete the process of disarmament and demobilization by 31 December 2005. In other words, the architects of the peace deal clearly understood that the disarmament and demobilization process constituted the most critical elements of the peace process that would determine the implementation of other terms of agreement.

The process of re-integration of GAM into Acehnese society is even addressed in a more comprehensive manner by the MoU. First, the GoI agrees to provide economic incentives and facilitation to former GAM combatants and pardoned political prisoners, including provisions for jobs. Second, the GoI promised to restore the political rights of former GAM members, including the granting of "amnesty to all persons who have participated in GAM activities",[10] the release of GAM prisoners and, more importantly, the right to political participation and to establish local political parties that meet national criteria. Third, former GAM combatants are also allowed "to seek employment in the organic police and organic military forces in Aceh without discrimination ..."[11]

On the institutional arrangement, the MoU stipulates the establishment of the Aceh Monitoring Mission (AMM) to ensure that the MoU is properly implemented. The AMM, which completed its task last year, was comprised of the European Union and ASEAN contributing countries (Brunei, Malaysia, the Philippines, Singapore and Thailand). It enjoyed an adequate mandate for carrying out its tasks, which included monitoring the DDR process, the human rights situation, and the process of legislative change. More importantly, the AMM was given full authority to rule on disputed cases and alleged violations of the MoU. Both the GoI and GAM agreed that the ruling of the Head of the AMM on disputed cases would be binding for both parties. Indeed, in practice, the AMM was able to demonstrate its impartiality whenever dispute arose.

The MoU clearly reflects a strong willingness from both sides to make compromises. In addition to the terms of agreement mentioned above, there are four other compromises that need to be mentioned specifically. The first relates to the question of Aceh's final status within the Republic of Indonesia. The provisions contained in the MoU provide for a federal-like arrangement in the relationship between Aceh and the rest of the country. This form of relationship clearly serves as a compromise between GAM's demand for independence, on the one hand, and the existing special autonomy offered by the GoI, on the other. The second compromise relates to the question of GAM's transformation and political participation. It serves as a compromise between the GoI's earlier demand for GAM's disbandment and the assertion by GAM leadership abroad that it represented a government-in-exile for Aceh. The third is the compromise on the agreed force level for the TNI troops to remain in Aceh, with its main responsibility to uphold external defence of the province.[12] The fourth is the unwritten agreement between GoI and GAM to set aside the problem of past abuses of human rights, even though the MoU stipulates that a human rights court is to be established in Aceh. This allows both sides to shelve the issue that could otherwise exacerbate the feeling of hostility and vengeance at the beginning of the peace process.

Finally, the MoU is creative, because it provides a formula that enables difficult issues to subside, if not disappear, if the implementation process proceeds smoothly and is completed. Two points are important in this regard. First, the MoU does not explicitly state that GAM should disband itself. Second, there is no explicit mention in the MoU text that GAM has officially dropped its demand for independence. Even though the absence of these two issues from the MoU has invited criticism from within Indonesia, it in fact serves two main purposes. While it provided a way for both sides to avoid sensitive issues that could have resulted in a deadlock during the negotiation process, the absence of these two issues would also reduce the possibility for the birth of splinter groups from within GAM. Moreover, the completion of the DDR programme will in effect amount to the de facto disbandment of GAM as an armed separatist movement.[13] Any declaration by GAM leadership to disband the movement and officially drop the demand for independence would undermine unity among its ranks, and might provide a pretext for hardline groups within the movement to mount a serious political challenge.

CHALLENGES OF POST-CONFLICT PEACE-BUILDING

It is important to note that what has been achieved so far is only the beginning of a long process of sustaining peace in Aceh. Whether a truly lasting peace in Aceh is indeed possible will depend on many other factors. The most important issue in this regard is the commitment and the ability of all stakeholders of peace — the central government, Aceh administration, former GAM members, political and religious leaders, civil society, the business community and ordinary citizens — to engage in a long-term and meaningful post-conflict peace-building process.[14] The key task in this regard is to create the conditions that would prevent conflict from re-emerging. For Aceh, the key challenges in the post-conflict peace-building efforts are immense. First, it is absolutely important to implement the unfinished key provisions contained in the MoU. Second, this is necessary for creating a mechanism for the peaceful settlement of differences and conflict. Third, a self-sustaining and durable peace can only be guaranteed by conscious efforts to address the root causes which gave rise to the conflict in the first place.

Unfinished Agenda of MoU

The first challenge is the completion of the re-integration process of former combatants into society. It had been mentioned earlier that one of the strengths of the Helsinki peace accord lies in the use of the DDR framework as an integral part of the peace process. In this regard, the disarmament and demobilization parts of the DDR have been finalized. By December 2005, for example, GAM completed its obligation to hand over 840 units of weapons as stipulated in the MoU. The GoI has also fulfilled its promise to withdraw security forces from the province, bringing the total number withdrawn to 31,681.[15] The reintegration process, however, has not been completed and is still an ongoing process. It is true that acceptance into society of the returning former combatants has been very high.[16] However, the re-integration process also requires provisions for economic facilitation and incentives for former combatants and the right of employment.[17] In this regard, the World Bank observes that "there is growing dissatisfaction and frustration at the pace of reintegration assistance [among former combatants]".[18] It has been noted also that "most of the 3,000 GAM fighters who surrendered their weapons last year to peace monitors are unemployed, and donor-backed reintegration has bogged down in squabbles over who gets priority, and how the money is disbursed".[19] This particular problem needs to be addressed, because lack of support for former combatants is a serious problem that could affect the peace process.[20]

Second, there is still the challenge of implementing the remaining key elements of the MoU. One key issue in this regard is the establishment of a Commission for Truth and Reconciliation (KKR) and a Human Rights Court.[21] As mentioned earlier, this will prove to be a contentious issue for peace-building efforts, because it might serve as a source of tension in intra-community relations and between Aceh and the central government in Jakarta, especially with the Indonesian Armed Forces (TNI) and the National Police (POLRI). Indeed, it has been noted that "while [the establishment of the Commission of Truth and Reconciliation] might assuage local grievances in the short term, the experiences of South Africa, Central Africa, the Balkans and Timor Leste suggest that such processes can do more harm to the social fabric than good".[22] Therefore, both the central government and the Aceh government need to tread carefully in addressing this sensitive issue. In other words, the implementation of transitional justice should also take into account the larger goal of sustaining peace in Aceh.

Third, the other remaining task is the fulfilment of the commitment to rehabilitate public and private property destroyed or damaged during the conflict.[23] It has been observed, however, that "dissatisfaction is rising amongst conflict-affected communities who have received very little but stand by watching tsunami-affected communities receive projects and support".[24] The gap between tsunami-related development projects and the lack of appropriate measures to address the grievances of conflict-affected communities has created new tensions within society, and between society and the government. Therefore, support for the conflict-affected communities is critical to the success of the peace process and should be made an integral part of the peace-building efforts. The use of the community-driven World Bank's Kecamatan Development Programme (KDP) by the Badan Reintegrasi Aceh (Aceh Reintegration Agency, or BRA) to address this particular problem is a good start.

Mechanism for Peaceful Conflict Resolution: The Imperative of Democratic Order

It was mentioned earlier that one key element of post-conflict peace-building is the creation of conditions that would prevent the recurrence of the conflict in the future. Within the Acehnese context, the deep-rooted conflict in the province was caused, among others, by the absence of democratic mechanisms for conflict resolution. I have argued elsewhere that people's resentments against the government, mostly expressed in the form of grievances over economic exploitation by the central government, had always been met with military oppression by Indonesia's New Order government.[25] Consequently, the long and bloody separatist war in the province reflected the preference for the use of violence as an instrument of dispute settlement in the absence of democratic mechanisms. In this regard, post-conflict peace-building in Aceh requires a deliberate effort by both central and local governments to create various mechanisms for peaceful conflict resolution within a democratic order. This task encompasses the need to strengthen the justice system, consolidate conflict management institutions (including informal and traditional ones), forge a harmonious central-regional relationship and create a democratic political order.

First, with regard to the need for a strong and viable justice system, the key challenge facing Aceh concerns the imperative for building effective, independent and impartial law enforcement agencies and legal institutions (the police, court and prosecutor's office). The strengthening of legal institutions and professionalism within the law enforcement agencies (especially within the police) requires a broader focus on security sector reform. While the reform of these two areas would be dependent upon broader reform at the national level, the specific context of Aceh as a post-conflict area should be taken into account. Good governance in legal and law enforcement agencies would certainly reduce the possibility for the growing sense of injustice. The importance of addressing this issue has become even more pressing due to the growing incidents of either local conflict or crimes in the province over the last few months.[26] One particular issue to be resolved in this regard is the division of labour among law enforcement agencies in Aceh.[27]

Second, even though the peace process is still on the right track, potentials for conflict have not disappeared. It has been noted that a variety of cleavages from the thirty-year conflict continue to manifest themselves in violent incidents and tensions that are especially related to aid distribution disputes and re-integration processes.[28] Preventing these types of tensions from becoming violent requires an effective dispute settlement mechanism. It is important to note, however, that dispute and conflict can be addressed outside formal legal institutions. It is important to strengthen traditional and custom-based institutions as informal managers of peaceful dispute settlements. Past experiences, even during the height of the conflict, clearly demonstrated that traditional and religious leaders were capable of playing a mediating role in settling conflicts. Equally important is the role of civil society organizations in managing the resolution of disputes within society. Capacity-building in alternative dispute settlement would encourage a more active participation of civil society organizations in peace-building efforts.

Third, during the New Order period, conflict in Aceh reflected a troubled relationship between the central government and Aceh. Indeed, one of the key features of New Order Indonesia was the excessive desire on the part of the central government to exercise absolute control over the regions. Jakarta, obsessed with the notion of national unity (*persatuan nasional*), imposed uniformity across the country without any regard for the nature of Indonesia as a pluralistic nation. Three decades of conflict in Aceh constituted a direct manifestation of the region's resistance to such desires by the centre. During the conflict, the credibility of the central government in Aceh was close to zero.[29] Since the signing of the peace accord in August 2005, Jakarta-Aceh relations continue to be precarious, as trust-building is still a work in progress. The key challenge facing both central and local government in this regard is the need to create communication in the spirit of mutual trust. In order to achieve this, it is necessary to narrow the "communication gap" identified by Aspinall as existing between most national policy-makers and actors in Aceh.[30]

Finally, peaceful resolution of disputes and conflicts, both within Aceh and between the centre and the region, requires a workable, democratic political order in Aceh. A free party system, fair and free elections, recognition of and respect for the role of civil society are all necessary but not sufficient ingredients for democracy. A democratic political order in Aceh will also require a strong commitment to respect the freedom of speech. All elements of Aceh's local government, including religious institutions, should not fall into the New Order's habit of suppressing the freedom of expression and limiting the rights of the people. Religious institutions, which have increasingly become central in Aceh's political structure, should also subscribe to democratic norms. After all, the MoU clearly stipulates that the province will have the right to draft "the legal code for Aceh on the basis of the universal principles of human rights as provided for in the United Nations International Covenants on Civil and Political Rights and on Economic, Social and Cultural Rights".[31]

Addressing the Root Causes: Post-Tsunami Reconstruction, Economic Development and Governance

The most difficult challenge for any post-conflict peace-building is the task of eliminating the root causes that brought about the conflict in the first place. Here, the important key driver of the conflict in Aceh was the lack of economic development in the province. Aceh consistently has been one of the poorest provinces in Indonesia, despite its abundant natural resources. The tsunami plunged the province into even deeper poverty. Coupled with decades of bad governance, the problem has become structural. The success of peace-building efforts and the future of peace in Aceh, therefore, would also depend on the ability of the government to deliver the promise of post-tsunami reconstruction, the acceleration of economic development, and the imperative of good governance.

First, the task of reconstruction and rehabilitation in Aceh was the main responsibility of the Badan Rekonstruksi dan Rehabilitasi (BRR). The BRR, despite the gigantic challenge it faces in rebuilding Aceh, was subject to both criticism and praise. The main complaint was directed at the slow pace of reconstruction and rehabilitation efforts. Speaker of the House Agung Laksono, for example, expressed his dismay in mid-2006 at the slow pace of reconstruction in Aceh.[32] A survey by Lembaga Survei Indonesia (LSI) found that, even though many Acehnese agreed that the BRR has done a far better job in 2006 than in 2005, the majority still complained that the pace continued to be slow.[33] The BRR also came under fire for alleged corruption in a number of development projects, prompting the attorney general's office to promise to look into the matter. Therefore, it is crucial for the BRA (successor to the BRR) to improve its performance. A failure to deliver the reconstruction promises would create new tensions, both within Aceh and between Aceh and the central government in Jakarta.

Second, a much more relevant issue for peace-building than post-tsunami reconstruction is the imperative of economic development. The lack of economic development, and the attendant feeling that Aceh was exploited by Jakarta, served as an important source of conflict in the province. Indeed, the social and economic condition in Aceh since 1998 had not been much different from that of the 1980s and 1990s, if not worse. In 1990, for example, Aceh contributed 3.6 per cent to Indonesia's gross domestic product. In 2001, this figure declined to just 2.2 per cent as a result of a significant decrease in the contributions from oil fields, agriculture and the processing sectors.[34] Poverty is the real problem in the province. According to former governor Abdullah Puteh, approximately 40 per cent of the province's 4.2 million people (1,680,000 people) are living under the poverty line.[35] This figure shows a significant increase from only 425,600 people in 1996 and 886,809 people in 1999.[36] Living conditions have not improved over the years. It is estimated that about half of Aceh's population still have earth or wooden floors, no access to safe drinking water and no access to electricity.[37]

Indeed, despite its abundant natural resources, by 2002 Aceh was the poorest province in Sumatra and the second poorest in Indonesia.

After the tsunami, the economic situation in Aceh deteriorated sharply. Despite the relative small impact on the national economy, the damage the tsunami has inflicted upon the Acehnese economy is by no means small. The World Bank estimates that the disaster impacted private assets and revenues by around 78 per cent: housing, commerce, agriculture, fisheries and transport vehicles and services were badly destroyed (US$2.8 billion, or 63 per cent of total damage and losses). As most Acehnese earn their livelihoods from agriculture, fisheries and commerce (accounting for 40 per cent of GDP), economic development programmes should restore these sectors in addition to attracting investment to provide job opportunities for the Acehnese. The success of the local government in accelerating economic development, especially in providing jobs, would certainly reduce a major source of conflict within society.

Third, it has been noted that "governance problems at the local level are central to the conflict system in Aceh, because they tend to undermine the legitimacy of the very institutions which are needed to mediate between local communities and the national government, and because they impede programs intended to alleviate structural causes of the conflict".[38] This problem in the governance sector will be one of the most difficult tasks the new Aceh government needs to address. Two challenges are of paramount importance in this regard: the corruption and capacity of local government. With regards to governance capacity, it has been noted that "local institutional capacity to distribute the massive benefits of autonomy is extremely low". The dramatic increase in local government revenues, for example, has not translated into concrete development outcomes.[39] Therefore, the focus on combating corruption and strengthening the capacity-building of local government — both at provincial and district levels — is imperative.

CONCLUDING REMARKS: THE CHALLENGE OF SUSTAINING PEACE AND LESSONS FOR OTHERS

The challenge for Aceh is how to sustain the positive factors by handling the problems that could push the peace process to collapse. However, all in all, the chance of success is still greater than the chance for failure. Therefore, there are few tentative, important lessons to be drawn from the Aceh peace process that might be relevant for other conflicts in the region and beyond.

The first is the role and the imperative of third-party facilitation or mediation. The deep-seated animosity and the absence of trust between the two parties made it impossible for both sides to negotiate directly. The involvement of the third party in the negotiation process, either as facilitator or mediator, helped to build productive communication between the conflicting parties, develop confidence, direct their attention to possible points of agreement, encourage compromise, offer alternative solutions, forge a peace deal and verify compliance to the terms of the agreement. All these essential prerequisites for reaching a peace agreement would not be possible without third-party involvement. In an internal conflict between a government and a secessionist movement such as in Aceh, the role of an independent third party acceptable to both parties in the conflict is absolutely crucial, and that can only be provided by an external party such as a state or an NGO. Indeed, despite the failure in implementation, the Humanitarian Pause and the COHA agreements would not have been possible if not for the Henry Dunant Centre (HDC). Similarly, the Helsinki accord would not have been possible if not for the effective role played by the Crisis Management Initiative (CMI), led by former Finnish president Martti Ahtisaari.

Second, the MoU is unique because it reflected new thinking within the Indonesian state, which no longer defined the question of sovereignty in a rigid way; this paved the way for creativity in searching for arrangements that could resolve conflicts. Two points of agreement in the MoU clearly reflect this new thinking. The first was the agreement to grant a federal-like status and high degree of autonomy for Aceh that opened the space for the assertion of local identity. The second was the agreement to bestow the AMM, an external actor in an internal conflict, the authority to rule when dispute occurred between the main parties to the conflict. The Indonesian Government no longer saw these two arrangements — the devolution of power and external involvement — as eroding state sovereignty. On the contrary, the arrival of peace in Aceh will enhance the sovereignty of the Indonesian state.

Third, a peace process will have a better chance of success if parties to the conflict agree to focus on peace incentives and confidence- and trust-building first, rather than on cases of human rights abuses. In the case of the Aceh conflict, both sides understood that the human rights issue was very divisive, and taking them on early in the peace process would not help reduce the feeling of mutual animosity. More importantly, the human rights perpetrators — many of whom were in the Indonesian army as well as in GAM — would not see any incentive to support the peace process. When they feel that the peace initiative might pose a threat to them, such groups may become difficult spoilers to deal with. The Indonesian military has repeatedly warned that it would not welcome any investigations of past human rights abuses against its rank and file, and instead calls all parties to look to the future instead.

Fourth, openness to outside participation in the implementation process greatly strengthens the commitment of both parties to comply with the terms of agreement. As mentioned earlier, the presence of an impartial third party is critical for ensuring that parties to the conflict will comply with the peace agreement. The role of the third party is particularly essential in generating confidence and trust during the implementation of arms decommissioning programmes and the demobilization of troops. In the case of Aceh, such third-party involvement came in the form of "a monitoring mission" comprised of EU and ASEAN contributing countries, with adequate mandate and authority to implement the most critical elements of the peace process (DDR), and to rule on violations of the terms of agreement. The conflicting parties would not be in a position to perform these two critical tasks.

Fifth, the peace process is difficult without political support at the national level. This support, which serves as a constituency for peace, greatly strengthens the government to search for creative measures for resolving the conflict. In the context of the Aceh conflict, the collapse of two previous agreements can partly be attributed to the lack of political support from within Indonesia's domestic constituencies. Many groups outside Aceh were opposed to negotiating with GAM and favoured a military approach to resolve the conflict. In the post-tsunami period, however, the constituency for peaceful resolution to the conflict grew stronger. In light of the devastation and the suffering inflicted by the tsunami on the Acehnese, it soon became "politically incorrect" to oppose any peace attempts through negotiation.

These five lessons might not be relevant to all types of conflict in all parts of the world. Each conflict possesses its own unique characteristics. However, some aspects of the problems, especially in secessionist conflicts, might be similar. In that context, what implications of the peace process in Aceh might have on other conflicts in the region will certainly depend on whether or not these lessons are learned. Indeed, as demonstrated in the case of Aceh peace process, new creative approaches to conflict resolution are made possible by learning from past experience. Without such creativity, the prospect for peace in Aceh would have been bleak indeed.

Notes

1. World Bank, "Rebuilding a Better Aceh and Nias: Stocktaking of the Reconstruction Efforts", *Brief for the Coordination Forum Aceh and Nias* (CFAN), 2005, p. 52.

2. World Bank, "The Aceh Peace Agreement: How Far Have We Come?" August 2006, p. 5 <http://www.internal-displacement.org/8025708F004CE90B/%28httpDocuments%29/ A4F908861EDDDE6FC125724200546781/$file/aceh+peace+agreement_WB_Dec06.pdf>.

3. This has become complicated by the post-tsunami reconstruction efforts, in which there has been substantial posturing by GAM and former combatants for employment and contracts. This has created significant tension for NGOs and the BRR.

4. Terance W. Bigalke, "Aceh Two Years After: Physical and Spiritual Challenges Remain," *EWC Insights* (Honolulu: East-West Center, 2006) <www.eastwestcenter.org/stored/pdfs/ Insights00101.pdf>, p. 3.

5. Parts of the discussion in this section are taken from Rizal Sukma, "Resolving Aceh Conflict: The Helsinki Peace Agreement", a briefing paper prepared for the Henry Dunant Centre for Humanitarian Dialogue (HDC), 2005.

6. Tamara Renee Shie, *Disarming for Peace in Aceh: Lessons Learned* (Monterey, CA: Monterey Institute of International Studies, 2003).

7. *Memorandum of Understanding between the Government of the Republic of Indonesia and the Free Aceh Movement* (MoU), Helsinki, 15 August 2005, point 4.3.

8. MoU 4.5.

9. MoU 4.6.

10. MoU 3.1.1.

11. MoU 3.2.7.

12. MoU 4.11.

13. On this issue, see Rizal Sukma, "Masalah Eksistensi GAM Pasca MoU/(The Question of GAM Existence after the MoU)", *Kompas*, 9 September 2005.

14. In this chapter, peace-building refers to a process that facilitates the establishment of durable peace and tries to prevent the recurrence of violence by addressing root causes and effects of conflict through reconciliation, institution-building, and political as well as economic transformation. See Boutros Boutros-Ghali, *An Agenda for Peace* (New York: United Nations, 1995).

15. World Bank, "The Aceh Peace Agreement: How Far Have We Come?", December 2006, p. 1 <http://www.internaldisplacement.org/8025708F004CE90B/(httpDocuments)/A4F908861EDDDE6FC125724200546781/$file/aceh+peace+agreement_WB_Dec06.pdf> (accessed 19 May 2009).

16. Ibid., p. 2. See also World Bank, "GAM Reintegration Needs Assessment: Enhancing Peace through Community-level Development Programming" (Jakarta: World Bank, 2006).

17. MoU 3.2.3 and 3.2.5.

18. World Bank, "The Aceh Peace Agreement", p. 3.

19. Simon Montlake, "In Aceh, Building Peace Amid Building Pains", *Christian Science Monitor*, 28 December 2006 <http://www.csmonitor.com/2006/1228/p01s02-woap.htm> (accessed 19 May 2009).

20. World Bank, "The Aceh Peace Agreement", p. 4.

21. MoU 2.2. and 2.3.
22. Ben Hillman, "Aceh's Rebels' Turn to Ruling", *Far Eastern Economic Review*, 25 January 2007.
23. MoU 3.2.4.
24. World Bank, "The Aceh Peace Agreement", p. 4.
25. Rizal Sukma, "Security Operations in Aceh: Goals, Consequences, and Lessons", Policy Studies No. 3 (Washington, DC: East-West Center, 2004).
26. World Bank/Decentralization Support Facility, "Aceh Conflict Monitoring Update", 1–30 November 2006, p. 5.
27. International Crisis Group (ICG), "Islamic Law and Criminal Justice in Aceh", Asia Report No. 117 (Jakarta/Brussels: ICG, 2006), p. 7.
28. World Bank/Decentralization Support Facility, "Aceh ConflictMonitoring Update", p. 1.
29. ICG, "Aceh: Can Autonomy Stem the Conflict?" Asia Report No. 18 (Jakarta and Brussels: ICG, 2001), p. 19.
30. Edward Aspinall, "Aceh/Indonesia: Conflict Analysis and Options for Systemic Conflict Transformation", paper prepared for the Berghof Foundation for Peace Support, August 2005, p. 13.
31. MoU 1.4.2.
32. M. Taufiqurrahman, "House Dismayed by Slow Aceh Reconstruction", *The Jakarta Post*, 14 June 2006.
33. Reuters, "Aceh Reconstruction Agency Seen Performing Better-Poll", *Alertnet*, 22 December 2006 <http://www.alertnet.org> (accessed 19 May, 2009)
34. Martin Panggabean, "War in Aceh: Its Economic Impact," *ISEAS Viewpoint*, 20 May 2003.
35. *Antara News Agency*, 28 October. 2003.
36. *Jakarta Post Online* <http://www.thejakartapost.com/special/0s_7_facts.asp> (no longer available as of 19 May 2009).
37. The World Bank, "Promoting Peaceful Development in Aceh", informal background paper prepared for the preparatory conference on Peace and Reconstruction in Aceh, Tokyo, 3 December 2002.
38. Aspinall, "Aceh/Indonesia: Conflict Analysis," p. 22.
39. Patrick Barron and Samuel Clark, "Decentralizing Inequality: Center-Periphery Relations, Local Governance and Conflict in Aceh", Social Development Papers, Conflict Prevention and Reconstruction, Paper No. 39 (December 2006), p. 15.

INDEX

www.ingramcontent.com/pod-product-compliance
Lightning Source LLC
Chambersburg PA
CBHW020339270326
41926CB00007B/244